ROTARY DRILLING SERIES

Drilling Fluids, Mud Pumps, and Conditioning Equipment

Unit I, Lesson 7
First Edition

by Kate Van Dyke

Published by

PETROLEUM EXTENSION SERVICE
THE UNIVERSITY OF TEXAS AT AUSTIN
Division of Continuing & Innovative Education
Austin, Texas

Originally produced by

INTERNATIONAL ASSOCIATION
OF DRILLING CONTRACTORS
Houston, Texas
1998

Library of Congress Cataloging-in-Publication Data

Van Dyke, Kate, 1951—
Drilling fluids, mud pumps, and conditioning equipment / by Kate Van Dyke. — 1st ed.
p. cm. — (Rotary drilling series ; unit 1, lesson 7)
Rev. ed. of: Mud pumps and conditioning equipment and circulating systems. 1982.
"Second impression 2003"—T.p. verso.
ISBN 0-88698-181-6 (alk. paper)
1. Drilling muds. 2. Oil well drilling—Equipment and supplies. I. Van Dyke, Kate, 1951-Mud pumps and conditioning equipment and circulating systems. II. Title. III. Series.
TN 871.27.V37 1998
622'.3381—dc22

2003023539

First Edition published 1998
Fifth Impression 2013
Printed in the United States of America

Catalog no. 2.107101
ISBN 0-88698-181-6

Contents

Figures

Tables

Foreword

For many years, the Rotary Drilling Series has oriented new personnel and further assisted experienced hands in the rotary drilling industry. As the industry changes, so must the manuals in this series reflect those changes.

The revisions to both text and illustrations are extensive. In addition, the layout has been "modernized" to make the information easy to get; the study questions have been rewritten; and each major section has been summarized to provide a handy comprehension check for the student.

PETEX wishes to thank industry reviewers—and our readers—for invaluable assistance in the revision of the Rotary Drilling Series. On the PETEX staff, Deborah Caples designed the layout; Doris Dickey proofread innumerable versions; and Kathryn Roberts saw production through from idea to book.

Although every effort was made to ensure accuracy, this manual is intended to be only a training aid; thus, nothing in it should be construed as approval or disapproval of any specific product or practice.

Ron Baker

Acknowledgments

Special thanks to Ken Fischer, Vice-President of Member Services, International Association of Drilling Contractors, who reviewed this manual and secured other reviewers. Tom Thomas and Michel Moy of Sedco Forex Schlumberger provided invaluable suggestions on the content and language. Without their assistance, this book could not have been written. In addition, special thanks to Leslie Kell, who managed to interpret some difficult sketches and make them into excellent drawings, and to Brandt/Tuboscope, Derrick Equipment Corp., M-D Totco, Mission Fluid King, and National Oilwell for assistance in obtaining photos.

Units of Measurement

Throughout the world, two systems of measurement dominate: the English system and the metric system. Today, the United States is almost the only country that employs the English system.

The English system uses the pound as the unit of weight, the foot as the unit of length, and the gallon as the unit of capacity. In the English system, for example, 1 foot equals 12 inches, 1 yard equals 36 inches, and 1 mile equals 5,280 feet or 1,760 yards.

The metric system uses the gram as the unit of weight, the metre as the unit of length, and the litre as the unit of capacity. In the metric system, for example, 1 metre equals 10 decimetres, 100 centimetres, or 1,000 millimetres. A kilometre equals 1,000 metres. The metric system, unlike the English system, uses a base of 10; thus, it is easy to convert from one unit to another. To convert from one unit to another in the English system, you must memorize or look up the values.

In the late 1970s, the Eleventh General Conference on Weights and Measures described and adopted the Système International (SI) d'Unités. Conference participants based the SI system on the metric system and designed it as an international standard of measurement.

The *Rotary Drilling Series* gives both English and SI units. And because the SI system employs the British spelling of many of the terms, the book follows those spelling rules as well. The unit of length, for example, is *metre*, not *meter*. (Note, however, that the unit of weight is *gram*, not *gramme*.)

To aid U.S. readers in making and understanding the conversion to the SI system, we include the following table.

English-Units-to-SI-Units Conversion Factors

Quantity or Property	English Units	Multiply English Units By	To Obtain These SI Units
Length, depth, or height	inches (in.)	25.4	millimetres (mm)
		2.54	centimetres (cm)
	feet (ft)	0.3048	metres (m)
	yards (yd)	0.9144	metres (m)
	miles (mi)	1609.344	metres (m)
		1.61	kilometres (km)
Hole and pipe diameters, bit size	inches (in.)	25.4	millimetres (mm)
Drilling rate	feet per hour (ft/h)	0.3048	metres per hour (m/h)
Weight on bit	pounds (lb)	0.445	decanewtons (dN)
Nozzle size	32nds of an inch	0.8	millimetres (mm)
Volume	barrels (bbl)	0.159	cubic metres (m^3)
		159	litres (L)
	gallons per stroke (gal/stroke)	0.00379	cubic metres per stroke (m^3/stroke)
	ounces (oz)	29.57	millilitres (mL)
	cubic inches ($in.^3$)	16.387	cubic centimetres (cm^3)
	cubic feet (ft^3)	28.3169	litres (L)
		0.0283	cubic metres (m^3)
	quarts (qt)	0.9464	litres (L)
	gallons (gal)	3.7854	litres (L)
	gallons (gal)	0.00379	cubic metres (m^3)
	pounds per barrel (lb/bbl)	2.895	kilograms per cubic metre (kg/m^3)
	barrels per ton (bbl/tn)	0.175	cubic metres per tonne (m^3/t)
Pump output and flow rate	gallons per minute (gpm)	0.00379	cubic metres per minute (m^3/min)
	gallons per hour (gph)	0.00379	cubic metres per hour (m^3/h)
	barrels per stroke (bbl/stroke)	0.159	cubic metres per stroke (m^3/stroke)
	barrels per minute (bbl/min)	0.159	cubic metres per minute (m^3/min)
Pressure	pounds per square inch (psi)	6.895	kilopascals (kPa)
		0.006895	megapascals (MPa)
Temperature	degrees Fahrenheit (°F)	$\frac{°F - 32}{1.8}$	degrees Celsius (°C)
Thermal gradient	1°F per 60 feet	—	1°C per 33 metres
Mass (weight)	ounces (oz)	28.35	grams (g)
	pounds (lb)	453.59	grams (g)
		0.4536	kilograms (kg)
	tons (tn)	0.9072	tonnes (t)
	pounds per foot (lb/ft)	1.488	kilograms per metre (kg/m)
Mud weight	pounds per gallon (ppg)	119.82	kilograms per cubic metre (kg/m^3)
	pounds per cubic foot (lb/ft^3)	16.0	kilograms per cubic metre (kg/m^3)
Pressure gradient	pounds per square inch per foot (psi/ft)	22.621	kilopascals per metre (kPa/m)
Funnel viscosity	seconds per quart (s/qt)	1.057	seconds per litre (s/L)
Yield point	pounds per 100 square feet (lb/100 ft^2)	0.48	pascals (Pa)
Gel strength	pounds per 100 square feet (lb/100 ft^2)	0.48	pascals (Pa)
Filter cake thickness	32nds of an inch	0.8	millimetres (mm)
Power	horsepower (hp)	0.75	kilowatts (kW)
Area	square inches ($in.^2$)	6.45	square centimetres (cm^2)
	square feet (ft^2)	0.0929	square metres (m^2)
	square yards (yd^2)	0.8361	square metres (m^2)
	square miles (mi^2)	2.59	square kilometres (km^2)
	acre (ac)	0.40	hectare (ha)
Drilling line wear	ton-miles (tn•mi)	14.317	megajoules (MJ)
		1.459	tonne-kilometres (t•km)
Torque	foot-pounds (ft•lb)	1.3558	newton metres (N•m)

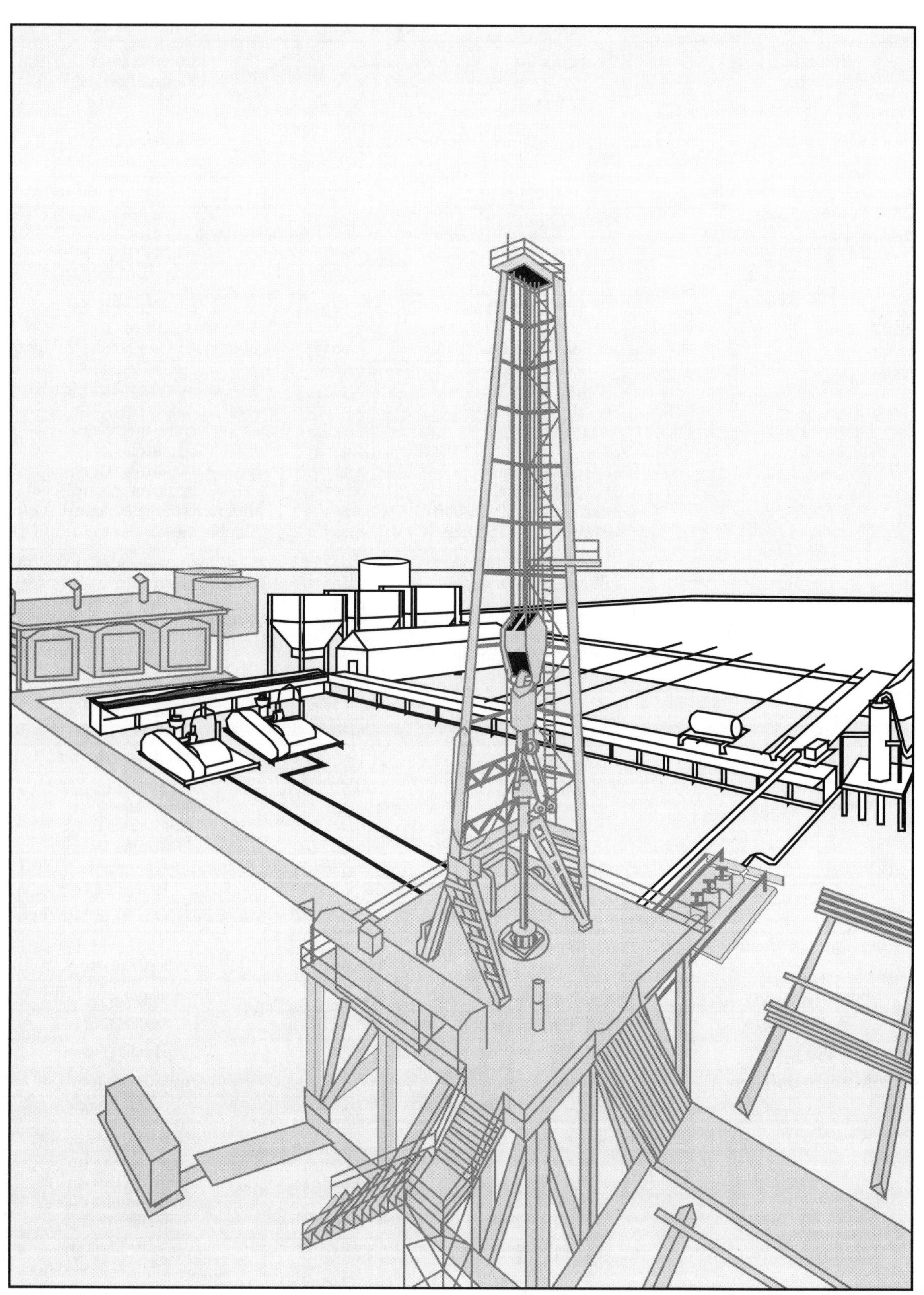

Introduction

A rotary drilling operation uses three systems that work together to make hole:

1. a rotating system that turns the bit,
2. a hoisting system that raises and lowers the drill stem and adds weight to the bit, and
3. a circulating system that moves a fluid down the drill stem, out of the bit, and back up the hole to the surface.

Drilling Fluids and Equipment

A typical circulating system on a rotary drilling rig includes the fluid that moves (called *drilling fluid*), equipment to move the fluid, and equipment to clean and *condition* the fluid. To condition the fluid means to circulate it through the circulating system so that all the additives in the fluid are distributed evenly throughout the system.

Drilling Fluids

A fluid is any substance that flows, so drilling fluid may be either a *liquid* or a gas. If in a liquid form, drilling fluid may be water or a mixture of water and oil with *additives*. Most oilfield workers call this mixture *drilling mud*. A gaseous drilling fluid may be either dry air or *natural gas*. Or it may be air or gas mixed with water and a foaming agent to form mist or foam.

About 95 percent of all drilling operations use fresh or *salt water* (water-base muds) as the drilling fluid. Water-base muds are relatively inexpensive to prepare because a water source is usually nearby, either as fresh groundwater or seawater. Oil muds are more expensive than water-base muds. Also they are harder to dispose of because of their possible detrimental effects on the environment. Air or gas drilling has advantages for certain formations, but its disadvantages limit its use.

Equipment for Circulating Drilling Mud

The main components of circulating equipment to move a liquid drilling fluid are the mud tanks, mud pumps, standpipe, rotary hose, swivel, drill stem and bit, and mud return line (fig. 1).

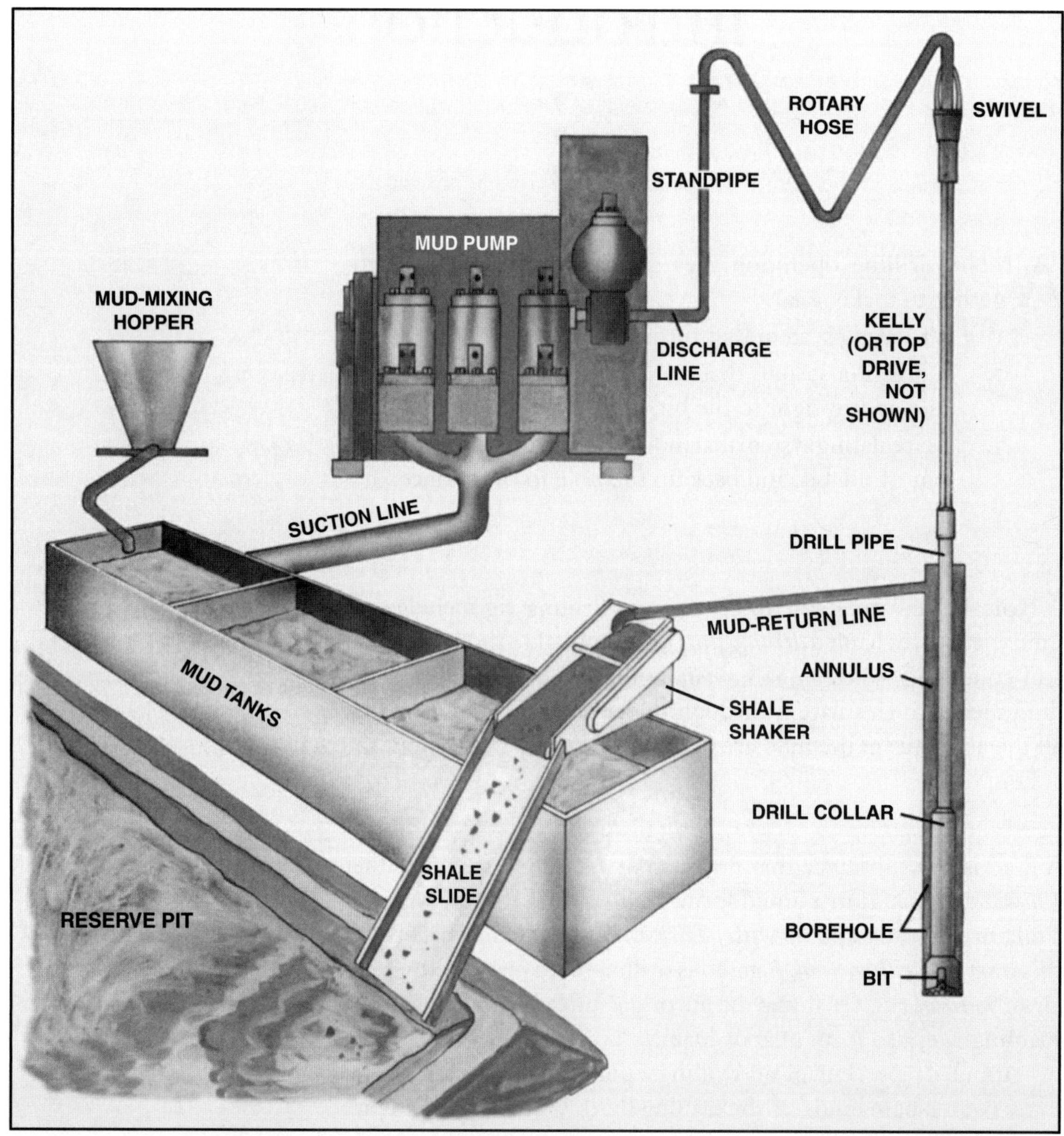

Figure 1. Drawn without the derrick, this diagram shows the relationship of the many components of the circulating system.

Mud tanks supply mud to the *pumps* and receive mud returned from the hole. Most rigs have two to four large main mud tanks. Crew members call them the *active tanks*. Usually, rigs also have several smaller tanks that can hold a portion of the mud for conditioning (treating) and reserve. Manufacturers typically make the tanks from steel. Mud tanks are normally rectangular in shape. They are open on top (they do not have a lid or a solid cover) but contractors customarily place open steel-mesh or fiber-grate walkways over the tanks to prevent crew members from falling into them.

Mud tanks also act as reservoirs where returned mud can cool off and slow down. Slow moving mud gives the cuttings time to settle out in sand traps. (*Sand traps* are small mud tanks into which sand in the mud falls.)

Large, heavy-duty pumps called *mud pumps* are the heart of the circulating system (fig. 2). Mud pumps have to be very powerful because they must be able to pump, at a high pressure, mud that can be heavy and thick. Today's most powerful pumps can handle as much as 2,200 input horsepower (over 1,500 kilowatts).

Figure 2. Modern mud pumps (Courtesy of Nabors Drilling USA)

To give you an idea of how much power this is, bear in mind that the most powerful automobile engines do not put out much more than 300 horsepower (210 kilowatts).

A mud pump forces drilling mud through the circulating system. Most land rigs have two pumps, even though the driller normally uses only one at a time during drilling. The second serves as a backup if the first requires repair. If, however, drilling requires a large *volume* of mud (for example, when the hole is large in diameter), the driller can use the two pumps together, compounding them. Offshore, where a drilling project may need very large volumes of mud, the rig may have three or, in a few cases, four mud pumps.

Mud pumps send the mud through the rotary hose, into the *swivel*, down the drill stem, and out of the bit. The mud returns to the surface through the *annular space*, or *annulus*. The annulus is the space between the walls of the hole and the drill string. Finally, it flows out through the mud return line back into the mud tanks.

Other Equipment

Before the used drilling mud flows back into the hole, it usually passes several pieces of special equipment. Crew members mount the equipment on the mud tanks to condition, or clean, the mud. The first piece of equipment is usually a vibrating screen, or sieve, called a *shale shaker* (fig. 3). The screen of the shaker separates the cuttings from the mud. The cuttings are larger than the openings in the shaker screen. The vibrating screen therefore carries the cuttings across the top of the screen and dumps them into the cuttings trough. The cuttings slide down the trough and into a *waste pit* for subsequent disposal. The mud flows through the shaker screen and returns to the circulating system. Most rigs use two or more shale shakers, each of which may have more than one screen. Modern *shakers* are sophisticated devices that are vital to cleaning the mud properly.

The mud often contains very small particles of rock, called *drilled solids*. Drilled solids are very small pieces of the formation that the mud carries up the hole with the cuttings. Because the drilled solids are so small, they fall through the shaker's screen openings along with the mud. Often, crew members call these drilled solids sand, because they often look like sand and are about the same size.

Figure 3. Cuttings carried by the mud are removed by two shale shakers on this location. (Courtesy of Nabors Drilling USA)

On most rigs the mud and the drilled solids or sand fall through the shale shakers into one or two sand traps. Typically the sand traps are tanks about 10 to 20 barrels (1.6 to 3.2 cubic metres) in capacity. They have sloping sides, which concentrate the drilled solids into a small area in the bottom of the tank. Concentrating the sand into a small area makes it easy for crew members to remove it.

The mud cannot stay in the sand traps very long because it has to move through the rest of the system and down the hole to do its work. Because there is not enough time to let all of the sand or solids settle, the contractor uses additional equipment to remove them. Generally, only solids larger than 74 microns in size have time to settle out in the sand traps. (A *micron* is very small, only one-millionth of a metre. A 74-micron solid is thus only about three-thousandths of an inch in size.)

Most drilling contractors install specialized solids control equipment to clean the finer drilled solids from the mud (fig. 4), such as a desander, desilter, mud cleaner, and centrifuge. If the mud contains entrained gas, the contractor will also install a degasser to remove it. (*Entrained gas* is gas that enters the mud as the bit drills a formation containing a small amount of gas.) What is more, the contractor may install a mud-gas separator to remove gas from mud that enters the *wellbore* during a *kick*. (A kick is the undesirable entry of formation fluids into the hole. Unlike entrained gas, a kick contains a large amount of gas and perhaps other fluids like salt water and oil.)

Other mud conditioning equipment includes *mud agitators*. Agitators keep the mud sufficiently stirred to prevent weighting materials from settling. Also, contractors usually install a small tank for mixing chemicals with water. In addition, they *rig up* a mud hopper for mixing dry material with the drilling fluid, and specialized instruments for monitoring the condition of the mud.

The area near the rig will also contain storage facilities for protecting bags of dry materials and chemicals from the weather, and possibly special containers for storing and handling dry materials in bulk form.

Figure 4. Additional circulating equipment includes a degasser, desilter, and desander, which are located over the mud pits downstream from the shaker.

The *derrickhand* on a rig is generally responsible for proper operation and maintenance of the mechanical equipment of the circulating system. Attending to the circulating fluid and the equipment of a circulating system makes up a large part of the daily work on a drilling rig. Crew members assist the derrickhand.

Circulating Air or Gas

A circulating system using air instead of a liquid drilling fluid needs compressors instead of pumps. Actually, air or gas is not circulated in the sense that it is used over and over again. Instead, it makes one *trip* from the compressors to the standpipe and swivel or *top drive*, down the drill stem, out the bit, and up the annulus to the surface. At the surface, the compressor blows the air or gas and cuttings (now ground to dust) out a *blooey line*, or vent pipe (fig. 5). Usually, crew members light a *flare* (a large gas flame) at the blooey line. The flare burns any gas coming out of the line to prevent it from collecting and possibly blowing up.

Before drilling with air or gas, the drilling contractor usually starts a well with water or mud as the circulating medium and sets the first string of casing. Most holes drilled with air or gas require changing back to drilling mud to finish drilling.

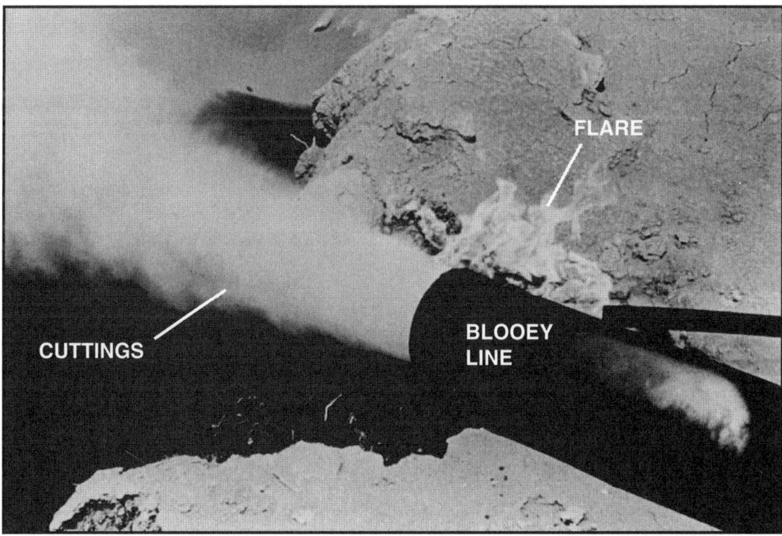

Figure 5. Cuttings are blasted out the blooey line. Should gas be encountered, the flare will ignite it so that the gas will burn away harmlessly.

History

The Lucas well, the famous gusher brought in at Spindletop near Beaumont, Texas in 1901, proved that rotary drilling was a good way to drill oil wells. Then as now, rotary drilling required a circulation system. That is, the rig had to circulate drilling fluid. In the case of Spindletop, the drillers circulated water with clay dissolved in it. They called it mud and the name has stuck to this day.

The Lucas well probably produced most of the mud it needed. The natural clay formations encountered in the well provided a passable drilling fluid when mixed with water. One legend is that the driller made the mud by driving cattle back and forth through a *pit* dug out of the ground and filled with water. Although this statement may be an exaggeration, it is quite likely that the drilling crew mixed the drilling mud they used in the circulating system in a nearby pit.

Circulating equipment for rotary drilling has improved over the years, but the circulation route remains essentially the same. In 1916, mud pumps were small steam-powered pumps (fig. 6) that produced *pressures* of 1,000 *pounds per square inch*, or *psi* (6,895 *kilopascals*, or *kPa*). Today's pumps, powered by the rig's *prime movers*, can move 200 to 1,200 gallons (757 to 4,542 litres) of liquid at pressures from 1,500 psi to 7,500 psi (10,342 to 51,712 kPa). What do these numbers mean? A hose carrying 200 gallons (757 litres) of water per minute at 1,800 psi (12,411 kPa) needs two firefighters to control it. A 3,000- to 4,000-psi (20,685- to 27,580-kPa) stream of water sent through a *nozzle* can drill through concrete.

Figure 6. Early mud pump

Drill bits have changed from early types, which were essentially just blades on a hollow shaft, to modern bits with jet nozzles to concentrate the pressure of the mud stream (fig. 7). Equipment for treating and cleaning mud is now much more efficient, to protect the investment that modern drilling muds have become. Since the 1920s, mud companies have formulated ever more sophisticated compositions of drilling mud that allow the driller more control under different drilling conditions. Supplying additives for mud is now a large industry. Mud companies maintain warehouse stocks near the principal oil fields and employ mud engineers to test the mud on jobs that use their material.

Operators began using air and gas for drilling in the early 1950s, and this method produced real improvement in penetration rates, but its disadvantages (see next chapter) have relegated it to a minor role.

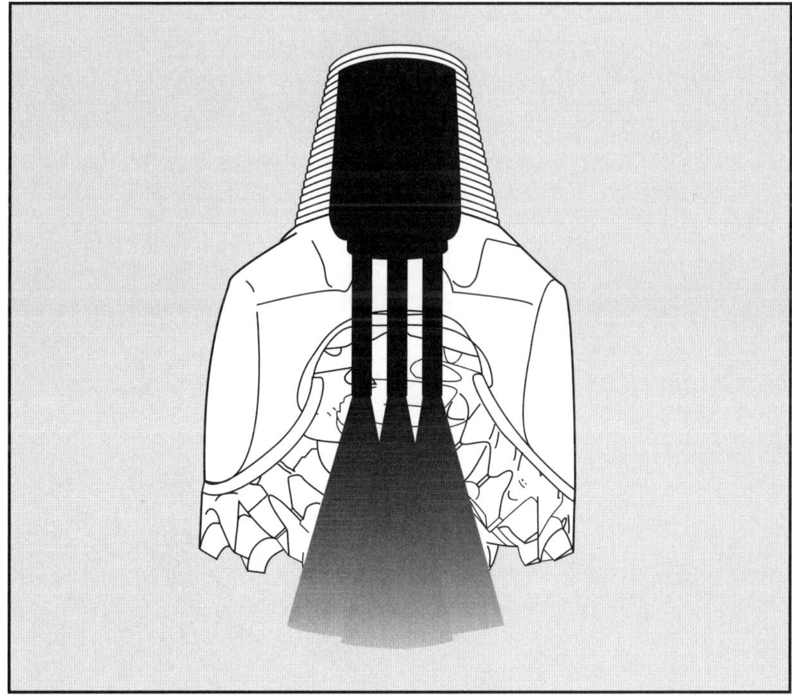

Figure 7. Watercourses in a roller cone bit

Circulating Systems

Drilling with a liquid and drilling with air or gas serve the same purposes, but each has advantages and disadvantages. The *operator* chooses the best system for a given drill site based on experience with previous wells in the same area.

Functions of the Circulating System

Circulating a fluid while drilling—

1. cleans the bottom of the hole;
2. transports cuttings to the surface;
3. cools the bit and lubricates the drill stem;
4. supports the walls of the wellbore;
5. prevents entry of formation fluids into the well, which can cause kicks or, if not controlled, leads to blowouts;
6. transmits hydraulic power to downhole equipment;
7. reveals the presence of oil, gas, or water that may enter the circulating fluid from a formation being drilled; and
8. reveals information about the formation, through the cuttings.

Not all types of drilling fluids handle these functions equally well. Air and gas have particular advantages and disadvantages compared to drilling muds. In the same way, not all muds are equal.

Cleaning the Bottom of the Hole

A bit must have a clean surface on which to work when making hole. For the bit to regrind the chips already broken off from the bottom of the hole is wasted effort. In addition, if drilling fluid does not sweep away the chips or cuttings as they form, the bit bogs down, and eventually the drill stem will not turn.

The drilling fluid exits the bit in a high-velocity stream. This stream blasts the bottom of the hole, and creates turbulence. The turbulence moves the chips away from the face of the formation as fast as the bit cuts them (fig. 8).

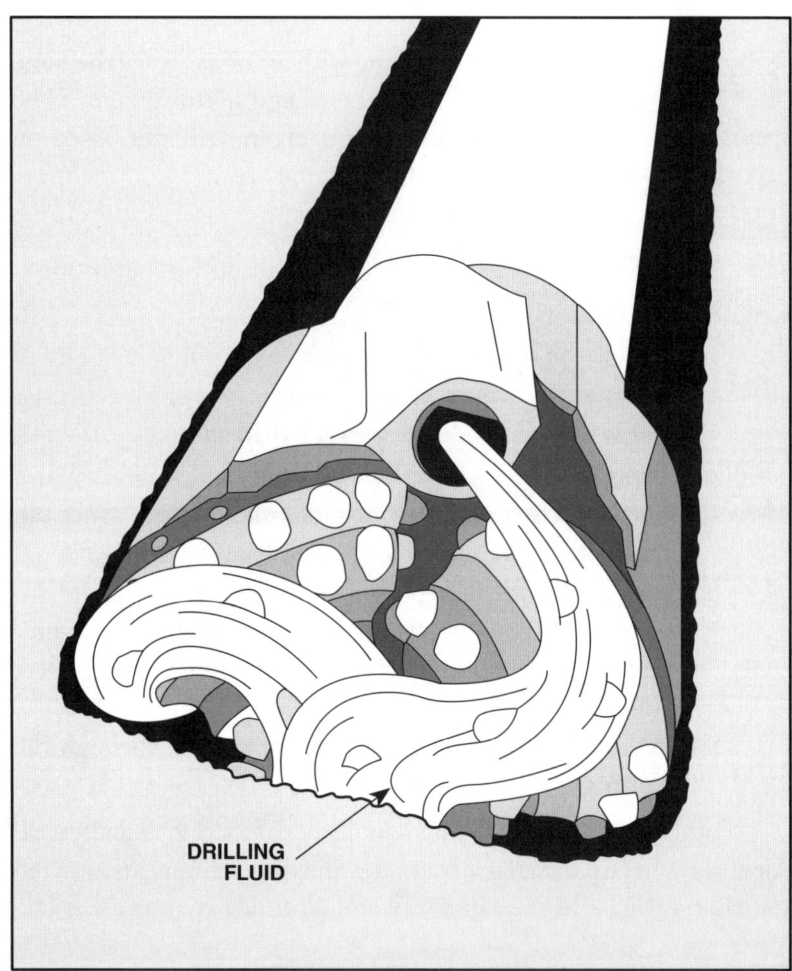

Figure 8. Circulating drilling fluid lifts cuttings.

Transporting Cuttings to the Surface

The drilling fluid carries rock chips, sand, or shale particles from the bottom of the hole as it moves up the annulus. In drilling with air or gas, the *annular velocity*, or speed at which the fluid flows up the annulus, determines how well cuttings move to the surface. Transport of cuttings when drilling with mud depends on annular velocity, viscosity, mud weight, yield point, and gel strength.

Annular Velocity

Liquid mud has to move from about 100 to 200 feet (30 to 60 metres) per minute to lift the cuttings and keep the hole clean. Put another way, mud has to have an annular velocity of about 100 to 200 feet (30 to 60 metres) per minute. Gas or air circulation requires an annular velocity of about 3,000 feet (915 metres) per minute. The only way air or gas can move cuttings to the surface is with pressure and velocity because a gas has no weight to hydraulically lift them. Air or gas drilling usually reduces the cuttings to dust by the time they reach the surface because they move very fast and pound against the tool joints and the wellbore on their way up.

Viscosity and Gel Strength

The *viscosity* of a drilling mud is its resistance to flow. For example, honey is more viscous than water. A viscous mud can transport more and heavier cuttings, so mud often contains a material that increases the viscosity, usually bentonite or polymers. (*Bentonite* is a naturally occurring clay that contains the mineral montmorillonite. A *polymer* is a chemical that consists of large molecules in repeating structural units.)

Bentonite, when added to freshwater mud, makes the mud *gel*—that is, it causes the mud to stiffen (solidify like gelatin) when circulation stops. This stiffening effect holds the cuttings in place within the mud instead of allowing them to fall to the bottom of the hole. When the mud starts moving again, the mud reliquefies and flows normally.

Gel strength is a measure of the mud's ability to suspend the cuttings. Viscosity and gel strength both help transport cuttings. However, a gel strength too high can be an undesirable property in a drilling fluid. Too high a gel strength can detrimentally affect surge and swab pressures in the circulating system.

Swab pressures occur when crew members pull the drill stem from the hole. It's like a syringe and needle when given a shot. The nurse sticks the needle into a bottle of vaccine and pulls the plunger on the syringe. The plunger then pulls, or swabs, vaccine into the syringe. Similarly, when crew members pull the drill stem from the hole, the pipe acts like a plunger and swabs formation fluids into the hole. If they swab too much fluid into the hole, the well kicks. It is easier for the crew to swab formation fluids into the hole with a high gel-strength mud than with a low gel-strength mud.

Surge pressures occur when crew members run the drill stem back into the hole. The opposite of swabbing, surging puts pressure on the walls of the hole. If the surge pressures are too high, they can break down (fracture) a formation. If the formation fractures, *lost circulation* can occur. That is, mud can flow into the fractured formation and be lost to it. Lost circulation can be bad because the mud's flowing into the formation reduces the height and thus the hydrostatic pressure of the column of mud. Reduced hydrostatic pressure can lead to a kick and possibly a blowout.

Cooling the Bit and Lubricating the Drill Stem

The combination of rotation and weight on the bit creates several hundred degrees of heat because of *friction*. Friction occurs inside the moving parts of the bit as well as outside the bit as it rotates against the formation. For example, the weight on an 8½-inch (216-millimetre) bit may be 60,000 pounds (26,700 decanewtons), about the weight of a railroad freight car. A 17-inch (432-millimetre) bit may require double that amount of weight. The rotation speed may be as fast as 4,240 revolutions per minute (rpm). Heat is the enemy of metal and especially of the diamonds in a diamond bit. Unless the circulating fluid removes the heat from friction, an expensive bit overheats and quickly wears out.

Oily substances contained in some drilling fluids can reduce friction in the bit bearings. Oil can also act as a lubricant between the drill stem and the walls of the hole. Air or gas circulation does a good job of cooling the bit because air or gas expands as it leaves the bit nozzles, which produces a cooling effect. Perhaps you have noticed that air released from an inflated tire or soccer ball feels cool. This is one reason bits last longer in air or gas drilling.

Supporting the Walls of the Well

A drilling mud with the proper characteristics can support a formation that might otherwise cave into a well. First, the weight of the fluid column in the hole creates pressure that pushes against the wall to support unconsolidated or loose formations that might fall or *slough* (pronounced "sluff") into the well (fig. 9). Second, this same pressure forces the liquid part of the mud a short distance into any porous and permeable formations the borehole has penetrated. The liquid goes into the formation leaving the solids in the mud on the wall of the hole, like plaster. This is *filter cake*, or *wall cake*. A good wall cake is a thin but solid sheath of material that keeps too much liquid from going into a porous and permeable zone.

Air and gas are not heavy enough to support the walls of the hole. For this reason, air or gas drilling is practical only in formations made of hard rock with little tendency to slough.

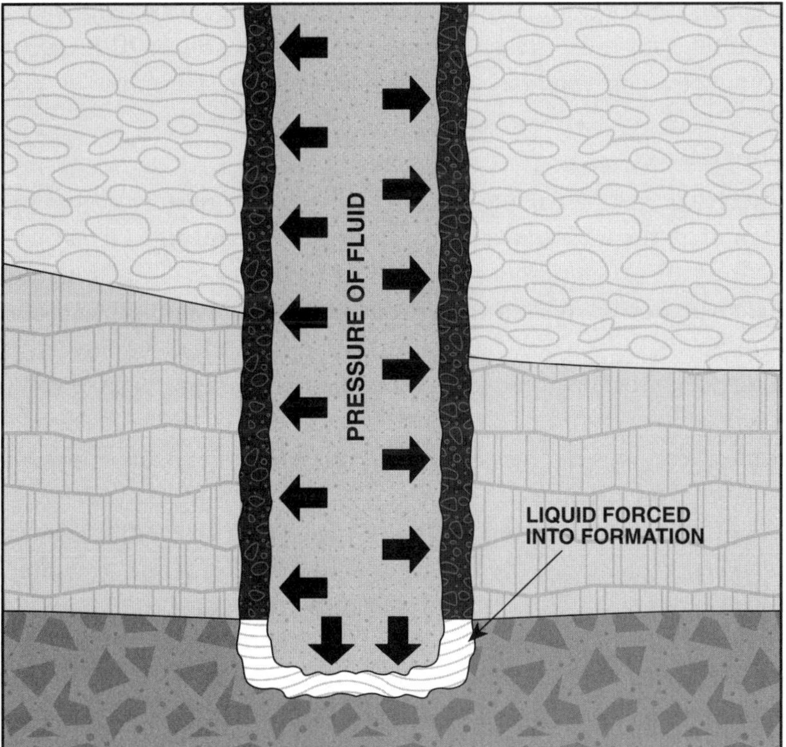

Figure 9. The drilling mud's hydrostatic pressure pushes on the sides and bottom of the hole, and forces the liquid part of the drilling mud into the formation.

Hydrostatic Pressure

Hydrostatic pressure is the force exerted by a fluid that is not moving. Hydrostatic pressure increases with the density and depth of the fluid. In a well, the *density*, or *weight*, of the fluid and the height of the fluid column in the hole determine its hydrostatic pressure. Both liquids and gases have hydrostatic pressure, but the hydrostatic pressure of a liquid is much greater. Drilling personnel usually express hydrostatic pressure in pounds per square inch (psi) or kilopascals (kPa).

The crew controls the hydrostatic pressure of drilling mud by increasing its density, or *weighting it up*. This weighted mud also does a better job of transporting cuttings.

If the mud weight is too high, however, its pressure can fracture the formation. Mud then flows into the fractures, instead of circulating back to the surface, causing lost circulation. Another problem that excessive mud weight can cause is a slower *rate of penetration (ROP)*. A pressure that is too high tends to hold the cuttings on the bottom so that they cannot easily move up the hole. As a result, the bit redrills a lot of cuttings instead of fresh formation, which slows the ROP. Pressure holding the chips on bottom is called *chip hold-down pressure*.

Filter Cake

Filter cake, also called mud cake or wall cake, is a plasterlike coating of mud on the walls of the hole. In a porous, permeable formation, hydrostatic pressure squeezes the liquid part of a drilling mud into the formation. If this filtration process stayed constant throughout drilling, the crew would have to add liquid to the mud continuously to make up for the liquid lost to the formation. However, the solid material in the mud is left behind as a filter cake (fig. 10). A good filter cake slows the loss of liquid, called *fluid loss*, from the mud to a very low rate.

Fluid loss that is too high is a problem for several reasons. One is that as long as the fluid filters into the formation, the filter cake keeps getting thicker. It may eventually become thick enough to reduce the diameter of the hole, causing tight spots where the drill string can get stuck. Second, muds with a high fluid loss may sometimes cause sloughing and caving of shale formations. Some shales are sensitive to water. That is, shale absorbs the water, swells up, and sloughs into the hole.

Finally, liquid entering the producing zones (the zones containing oil) may reduce the rate of oil flow when the well is ready to produce. This phenomenon happens because, like shales, some producing zones are sensitive to water. The water enters tiny openings of the zone and causes the surrounding rock to swell,

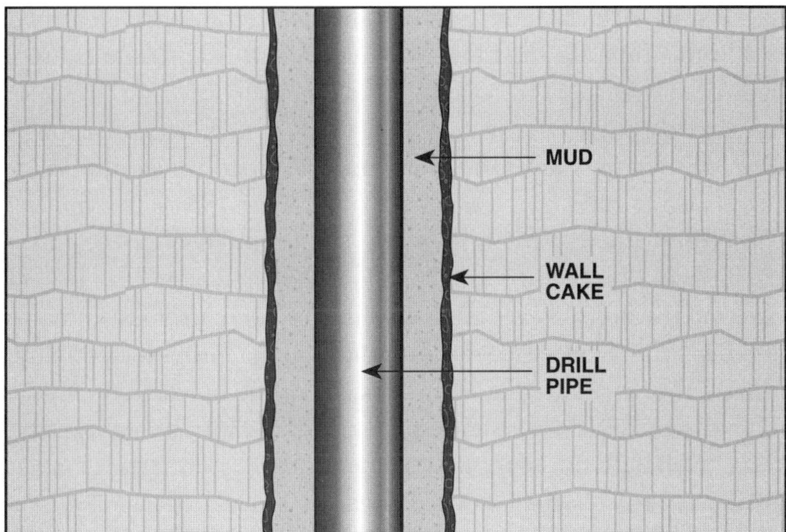

Figure 10. Solid particles in the drilling mud plaster the wall of the hole and form an impermeable wall cake.

thus blocking permeability. This blocked permeability is *formation damage*.

A good filter cake is thin, slick, and virtually impermeable. The crew may add finely ground clays or other substances such as polymers to drilling mud to improve its ability to form a filter cake, or *wall-building ability*.

Preventing Entry of Formation Fluids into the Well

Gas, oil, or water in the drilled formations also exert pressure. The hydrostatic pressure of the mud can be the same, greater, or less than the pressure in the drilled formations that contain fluids. When pressure in the wellbore is the same as the pressure in the formation, crew members say that the hole is balanced. When the pressure in the wellbore is greater than the pressure in the formation, they say that the hole is *overbalanced*. When formation pressure is greater than wellbore pressure, they say that the hole is *underbalanced*.

If formation pressure is greater than the hydrostatic pressure of the drilling fluid, formation fluid enters the well, causing a kick. A kick is a danger signal because, if not controlled quickly, it can lead to a blowout. All crew members need to recognize some of the obvious signs of a kick: (1) the volume of fluid returning from the hole increases; (2) when the driller shuts down the mud pump, mud continues to flow from the well; and (3) the level of mud in the mud tanks increases. Drillers keep an eye on instruments that measure mud flow and tank volume to watch for these conditions.

To prevent kicks, crew members should keep the hole full of mud at all times and use mud that is heavy enough to provide enough hydrostatic pressure to balance or overcome formation pressure. Adding weighting material to drilling mud can make it dense enough to hold back almost any formation pressure. To stop a kick, the driller closes special valves on the wellhead called *blowout preventers* (fig. 11). Closed blowout preventers hold *back-pressure* on the fluid column at the surface. This back-pressure prevents more formation fluids from entering the hole. To kill the kick, the crew weights up the mud to develop enough hydrostatic pressure to control formation pressure. Finally, they circulate the kick fluids out of the hole and the heavy mud into the hole to reestablish control of the formation pressure. Once the hole is full of the new, heavy mud, they open the blowout preventers and continue normal operations.

Air and gas, because of their low hydrostatic pressure, are not a good choice in a well where the operator expects the formation pressure to be high.

Figure 11. Blowout preventer stack on a land rig. At the top of the stack is the annular preventer. Three ram preventers are nippled up below the annular. Note the vertical pipe running from top to bottom—this is the mousehole.

Powering Downhole Equipment

Some drilling operations use equipment in the hole that is not rotated by the top drive or the rotary table. For example, directional and horizontal drilling use a downhole motor to turn the bit. Oil company engineers and technicians have learned that they can produce some reservoirs better if they drill a horizontal (instead of a vertical) well through the reservoir. In *horizontal drilling*, crew members begin by drilling a normal, vertical hole to a given depth. This depth is usually high above the reservoir's depth. They then deflect the hole from vertical. Over a distance of perhaps hundreds of feet or metres, crew members curve the vertical hole until it becomes horizontal. The hole actually runs parallel to the surface rather than perpendicular to it.

When drilling a horizontal well, the rig crew cannot use a top drive or a conventional rotary table to rotate the drill string and bit. The reason has to do with the way in which they deflect the hole from vertical. Usually, crew members run a device with a slight bend in it—a *bent sub*. This bend, usually from 1 to 3 degrees, points the drill string off to one side of the vertical hole (fig. 12a). Crew members carefully point (orient) the bend in the string to make the bit go in the right direction.

While 1 to 3 degrees is not very much from vertical, the bend is enough to start the horizontal hole. Once started, crew members use special techniques to increase the curve more and more until the hole becomes horizontal.

The drill string now has a bend in it (drill pipe is flexible—it can bend a lot without breaking or permanently bending). The bend is also pointing the bit in the direction it needs to go. Crew members therefore cannot rotate the entire drill string. If they did, they would get a wobbling motion, rather than the straight rotating motion they get with a vertical string. Also, the bit would not drill in the right direction.

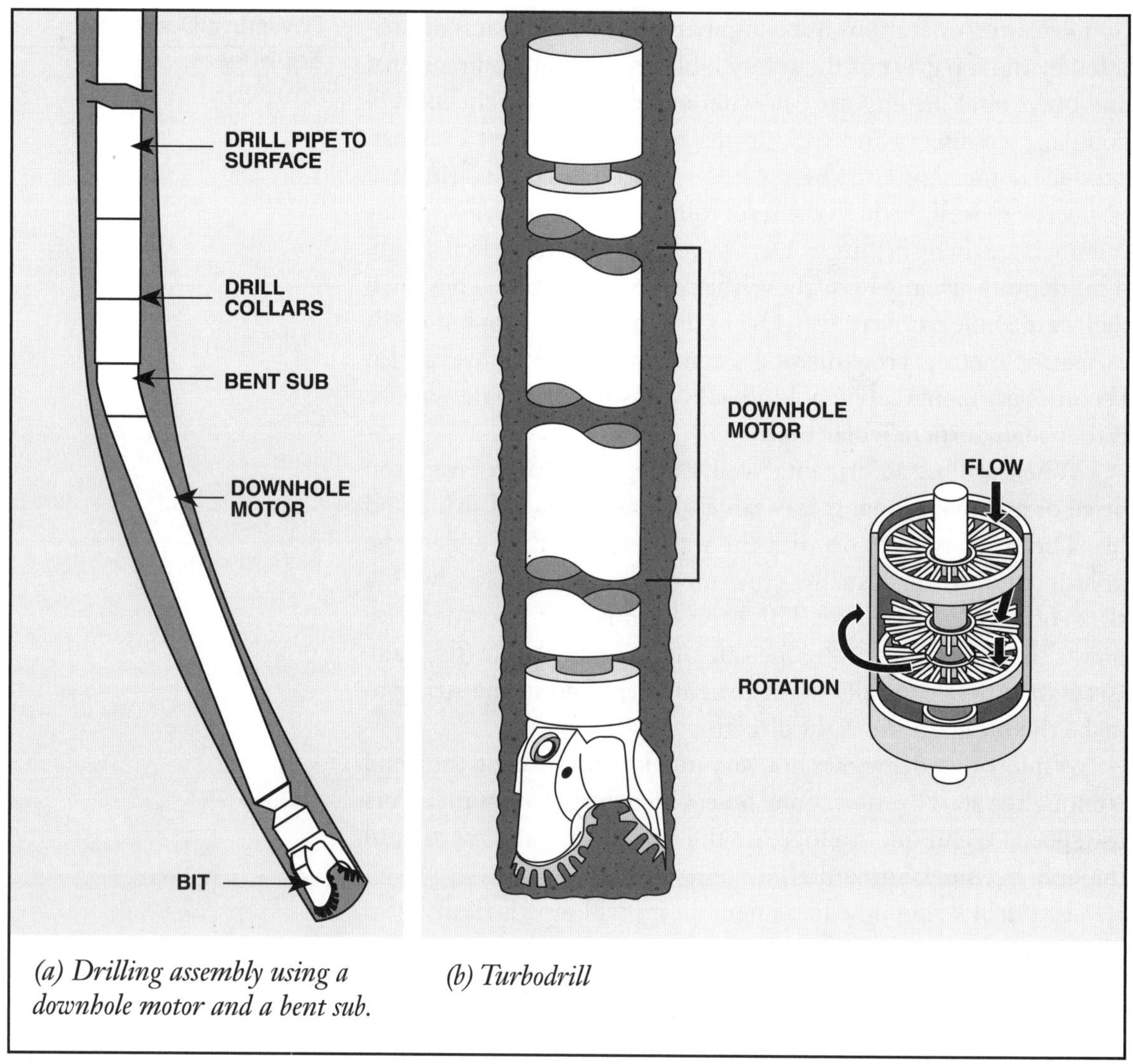

(a) Drilling assembly using a downhole motor and a bent sub.

(b) Turbodrill

Figure 12.

Since the rig cannot rotate the entire drill string when drilling a horizontal well, crew members install a *downhole motor* (sometimes called a mud motor). A downhole motor is a strong length of pipe, inside of which is a special device that rotates the bit (fig. 12b). Crew members attach the motor to the bottom of the drill string, and attach the bit to the bottom of the motor. With the bit and motor oriented in the right direction, the mud pump forces mud down the drill string and to the motor. In the motor, the mud strikes either a spiral-shaped shaft or several turbine blades. The great hydraulic force of the mud turns the shaft or blades. The shaft or blades (turbines) then turn the attached bit. Only the bit turns.

Getting Information about the Formation Rock and Fluids

A geologist at a drill site periodically examines the cuttings to determine what formations the hole has penetrated. Mud engineers test drilling mud to see how much water, oil, or gas is entering the wellbore. If the well is a *wildcat* (the first well drilled in an area), the operator needs as much information as possible to decide whether to continue drilling. When drilling in a known area, geologists still test the cuttings and fluid and compare the results with records from previous wells. Changes can reveal information about the shape and extent of the formations and the producing zones. For example, finding a particular type of rock a few feet or metres deeper than in a nearby well indicates that this layer of rock is sloping downward.

To summarize—

Functions of the circulating system

- To clean the bottom of the hole
- To transport cuttings to the surface
- To cool the bit and lubricate the drill stem
- To support the walls of the wellbore
- To prevent kicks and blowouts
- To transmit hydraulic power to downhole equipment
- To reveal the presence of oil, gas, or water that may enter the circulating fluid from a formation being drilled
- To reveal information about the formation, through the cuttings

---

Mud Circulating Systems

The path of fluid in a circulating system remains much the same whether the fluid is liquid or gaseous. The equipment, however, varies somewhat. A liquid drilling fluid requires tanks for storage and pumps to move it through the circulating path.

Equipment

The main components of a circulating system to move a liquid drilling fluid are the mud tanks, mud pumps, standpipe, rotary hose, swivel, drill string with bit, and mud return line (fig. 13).

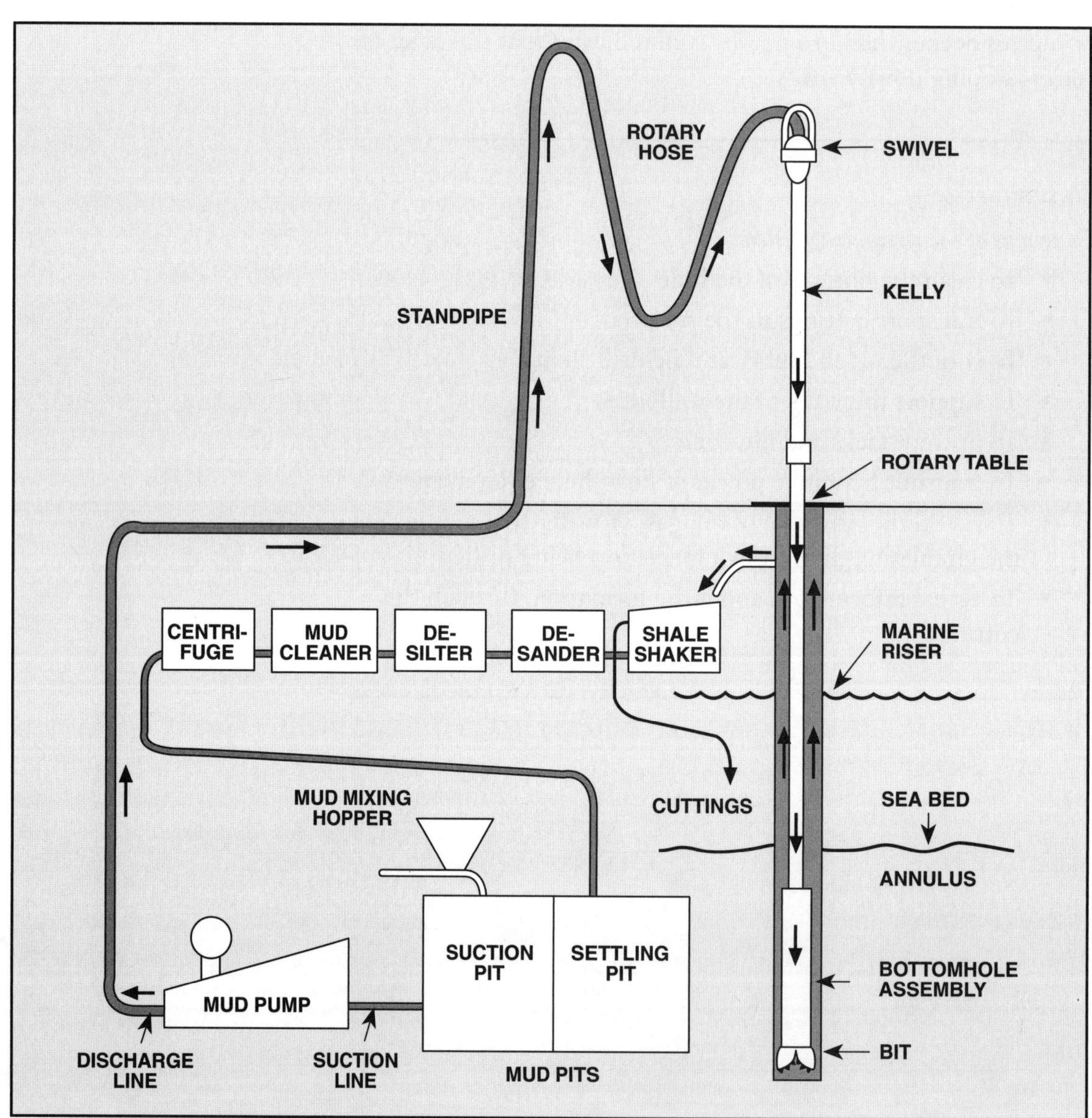

Figure 13. This schematic diagram shows the circulating path of drilling mud.

Mud Tanks

The *mud tanks* are the beginning and end of a mud circulation system. Their functions are to—

1. supply mud to the pump to begin circulation;
2. receive mud circulated out of the well;
3. store reserve mud to fill the hole when crew members remove the drill stem;
4. serve as a reservoir to allow the mud to cool and solids to settle out of the mud; and
5. serve as a reservoir for mixing additives into the mud.

Originally, drilling companies dug pits into the ground to hold the mud and called them *mud pits*. Steel tanks are now standard, although some oilfield workers still refer to the tanks as mud pits.

Any tank that is in use for circulaiton is an active tank. The tank that supplies mud to the pump is the *suction tank*. A *suction strainer* in the pump's suction line keeps items such as gloves and welding rods accidentally dropped into the suction tank from being drawn into the pump. If such items get into the pump valves, a shutdown is the certain result. A *reserve tank* serves mainly as a place to send an overflow of mud. It may also be used to store a batch of mud that is heavier than the circulating mud or to store mud for recycling.

Steel Tanks. A mud circulating system uses two, three, four, or even more steel tanks. Tanks are better than earthen pits for several reasons. Manufacturers make steel tanks with a known volume, which makes chemical treating of the mud easy to plan. Steel tanks are easy to clean. Most have cleanout gates or doors on the sump (low) side of the tank. What is more, the drilling contractor can arrange piping for mud flow in any configuration, making it possible to bypass one tank and send the mud directly to another tank as needed. Crew members can easily install auxiliary equipment, such as mud-cleaning equipment, mixers, and measuring instruments on steel tanks. Rig-up on new locations is faster when the crew can connect suction lines and piping directly to the mud tanks.

Mud tanks must be large enough to hold enough mud to fill the circulating system when the drill pipe is on bottom at total depth. They must also be large enough to provide an adequate settling area. A deeper hole needs larger and more tanks than a shallower hole.

Other Tanks. The mud system uses a number of smaller steel tanks as well as large mud tanks. One small but important tank, the *trip tank*, holds a small quantity of mud that crew members use when removing pipe from the hole. They pump mud out of the trip tank and into the hole to replace the volume the pipe took up while it was in the hole. A trip tank has a capacity of about 10 to 30 barrels (1.6 to 4.8 cubic metres). It also has a float gauge on the side that accurately measures the amount of mud in the tank. When tripping the drill string out of the hole, rig personnel closely watch the level of mud in the trip tank. They record the amount of mud that goes out of the trip tank and into the hole to replace the volume the drill stem took up. They use a *trip sheet* (fig. 14) to record the volumes. The hole should take the same amount of mud that the pipe occupied. Keeping the hole full of the right amount of mud ensures that the hydrostatic pressure of the mud remains high enough to offset formation pressure.

The trip tank is a critical piece of equipment because most kicks occur during trips. Filling out the trip sheet accurately is also crucial. Crew members must know whether the hole is taking the right amount of mud (called fill-up mud) to replace the drill pipe. If they swab in formation fluids during a trip, the hole will not take as much fill-up mud as it should. Put another way, too little fill-up mud indicates that the crew has swabbed formation fluids into the hole. Being aware that they have swabbed fluids into the hole allows the crew to take steps to prevent a kick and possible blowout.

Another small tank is a *slug tank*. A slug tank may contain a heavy mud that the crew uses when tripping out of the hole. The driller pumps the heavier *slug* of mud into the top of the drill pipe immediately before pulling the pipe. The heavy slug of mud pushes the level of mud already in the pipe down below the rig floor, so the pipe is empty as it comes out of the hole. The slug prevents mud from spilling onto the rig floor and crew members when they break out (unscrew) a length of pipe. Another small tank on some rigs is a *pill tank*. A pill tank may contain an especially viscous quantity of mud, such as one made up mostly of bentonite and water. The driller can pump the pill of viscous mud down the hole and *spot* (place) it at or near a formation that is causing lost circulation problems. The viscous pill may cure the problem. Or a pill tank may contain a special chemical that can help retrieve pipe that is stuck downhole. The chemical is very slick or slippery. When the driller spots the slick chemical at the point where the pipe is stuck, the chemical may help free the pipe. Rigs with only one small auxiliary tank use it for multiple purposes, such as slug, pill, or chemical spotting.

RIG: ____________________ DATE: ____________________

WELL: ____________________ TIME: ____________________

DRILLER: ____________________ **TRIP SHEET** DEPTH: ____________________

REASON FOR THE TRIP: ____________________

Number of stands to have top of DC's one DP stand below BOP's: ____________________

PULL ON:	✓
EVEN	
SINGLE	
DOUBLE	

DISPLACEMENT:	DC1	DC2	OTHER	HWDP	DP1	DP2
Size						
bbl/ft or						
bbl/stand						
x ft or stands						
= Vol. (bbls)						

STAND NO.	TRIP TANK GAUGE	CALCULATED Hole Fill (bbls) per Increment	MEASURED Hole Fill (bbls)		DISCREPANCY		REMARKS
			per Increm'	Accumul.	per Increm'	Accumul.	
0							

Figure 14. Trip sheet with spaces to show displacement of drill stem components, number of stands pulled, and calculated versus actual amounts of fill-up mud. (Courtesy of Sedco Forex Schlumberger)

Mud Pumps

A rig usually has at least two large mud pumps, also called *slush pumps*. Their function is to move huge quantities of mud from the mud tanks, down thousands of feet or metres of drill pipe, out small nozzles in the bit, up the annulus, and back to the mud tanks. This route requires powerful pumps that can supply a great amount of pressure to overcome the forces of friction and gravity. The prime mover on the rig powers the mud pumps. A large mud pump may operate at up to 2,200 horsepower (over 1,500 kilowatts), some seven times the power of a car.

This book deals with mud pumps in detail later.

Mud Manifold

A manifold is piping that connects two (or more) pumps (fig. 15). It divides the flow of mud into several parts, combines several

Figure 15. High-pressure mud manifold

flows into one, or reroutes a flow to different destinations. The *mud manifold*, or *pump manifold*, is an arrangement of valves and piping that permits a wide choice in the routing of suction and discharge fluids among two or more pumps. The pipes are connected by couplings that crew members can tighten and loosen with a hammer. Special valves allow them to isolate any part of the manifold. Being able to isolate parts of the manifold allows crew members to reroute the mud flow around a leak during a critical operation. It also allows them to repair one part of the system while the other part is operating under pressure.

Crew members usually install short lengths of flexible hose from the pumps to the manifold. The hose has special connectors that allow quick rig up. The flexibility of the hoses allows them to absorb some of the vibrating pulsations that originate in the pump. For this reason, these hoses are called *vibrator hoses* or *shock hoses*.

Standpipe and Rotary Hose

Circulation of drilling mud begins from the mud tanks. The mud pump takes in mud through a *suction line* from a suction tank (fig. 16). The mud pump pushes the mud through a series of pipes that direct the drilling fluid down the hole. First, the pump moves the

Figure 16. Suction line from mud suction tank to mud pump. (Courtesy of Nabors Drilling USA)

mud through a discharge line to the *standpipe* (fig. 17). The standpipe is a rigid pipe whose bottom end connects to a mud pump's discharge line and top end to the rotary hose. Clamps secure the standpipe firmly to the derrick or mast. A curved fitting known as a *gooseneck* on top connects the standpipe to one end of the flexible *rotary hose*, or *kelly hose*, the next tube in the mud's path down the

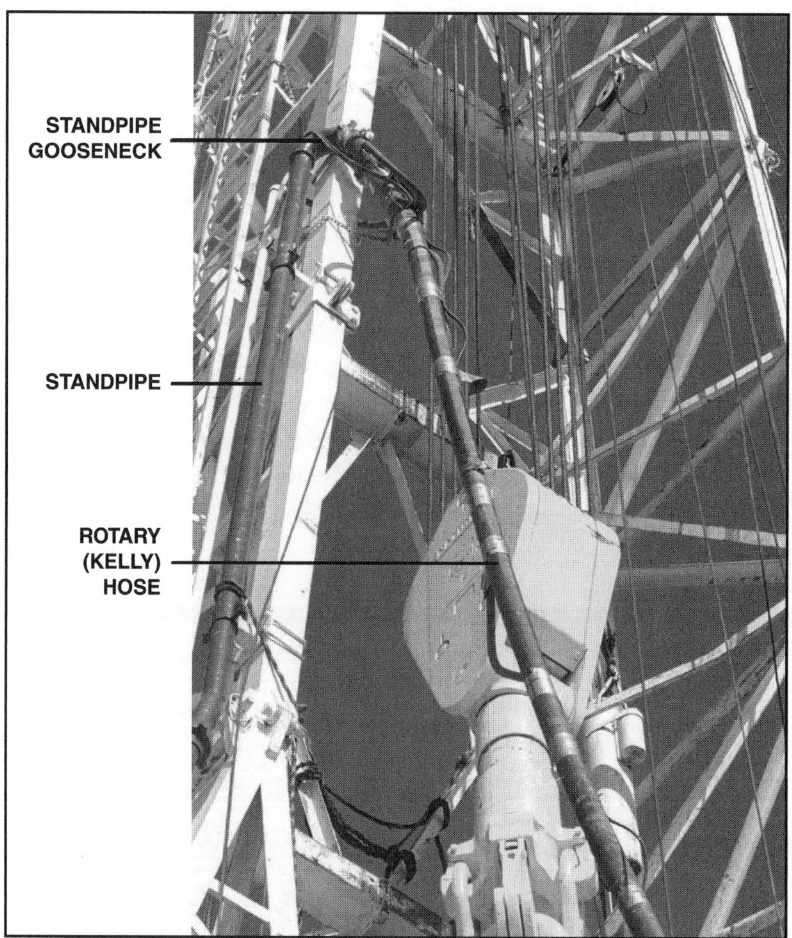

Figure 17. The mud pump sends mud up the standpipe and into the rotary hose. (Courtesy of Nabors Drilling USA)

hole. The rotary hose is flexible because it must move up and down with the drill stem as the driller hoists and lowers it. The other end of the hose is attached to a second gooseneck on the swivel or top drive (fig. 18). Crew members attach safety chains to clamps near the two ends of the hose. The chains prevent the hose from whipping around should one of the normal fittings fail.

Figure 18. Rotary hose attached to the swivel gooseneck

The standpipe keeps the rotary hose clear of the rig floor when the *kelly* or the first length of drill pipe has been drilled down and the swivel or top drive is near the rotary table (fig. 19). A typical configuration uses a 48-foot (15-metre) standpipe with a 55-foot (17-metre) hose. This permits the swivel and kelly or top drive to travel upward in the derrick about 80 feet (25 metres) above the rotary table.

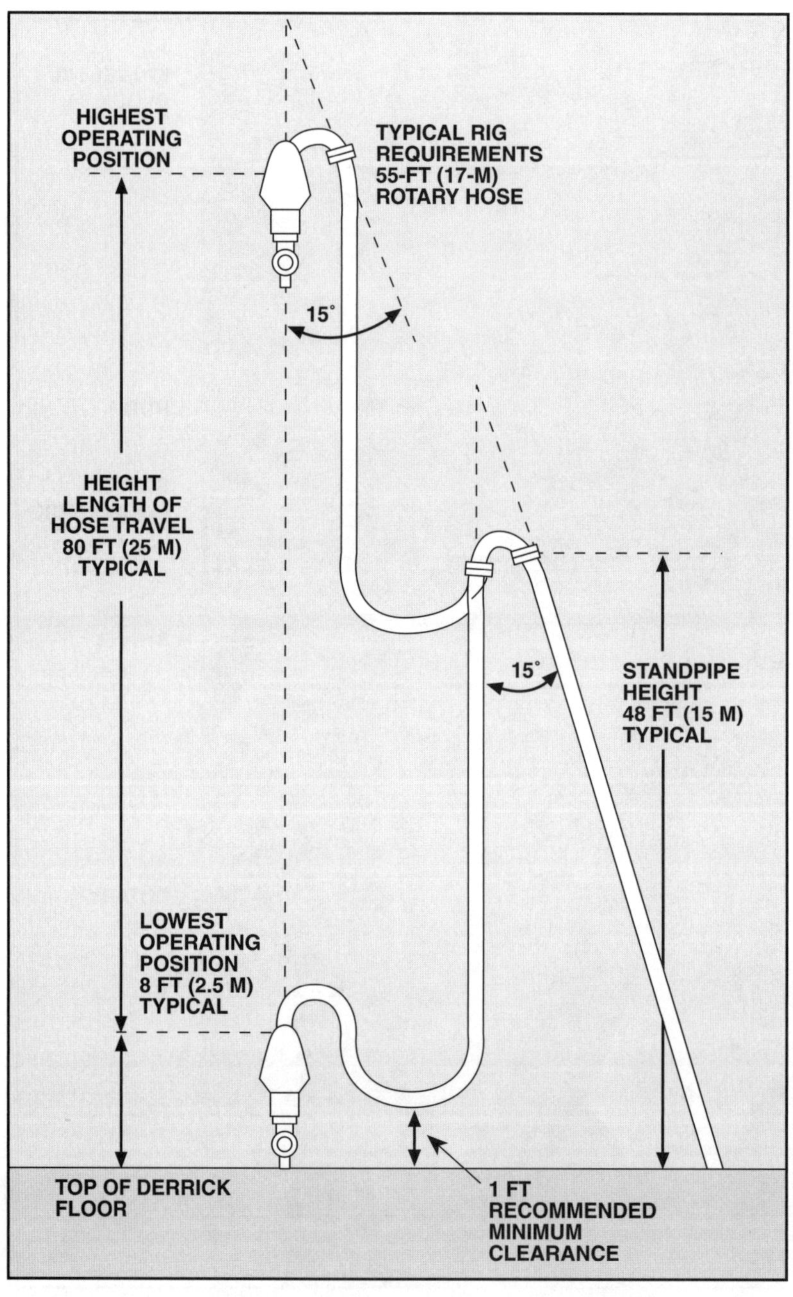

Figure 19. Standpipe height for a 55-foot (17-metre) hose

The crew needs this distance to connect 30-foot (9-metre) lengths of drill pipe in the mousehole with a 40-foot (12-metre) kelly, with enough of a margin to prevent the hose from bending too sharply at its connection with the standpipe.

The rotary hose must be strong enough so that it does not leak under high pressure and does not wear out from abrasive solids in drilling fluids. It is made of a special synthetic rubber reinforced with steel mesh (fig. 20). Manufacturers follow *API (American Petroleum Institute)* specifications that set guidelines for pressures that the hoses can withstand and their dimensions. API publishes these and specifications for other equipment in *API Specifications for Rotary Drilling Equipment*, known as *API Spec.* 7. (This and other API publications may be purchased from API, Order Desk, 1220 L Street, N.W., Washington, D.C. 20005; telephone 1-202-682-8375.)

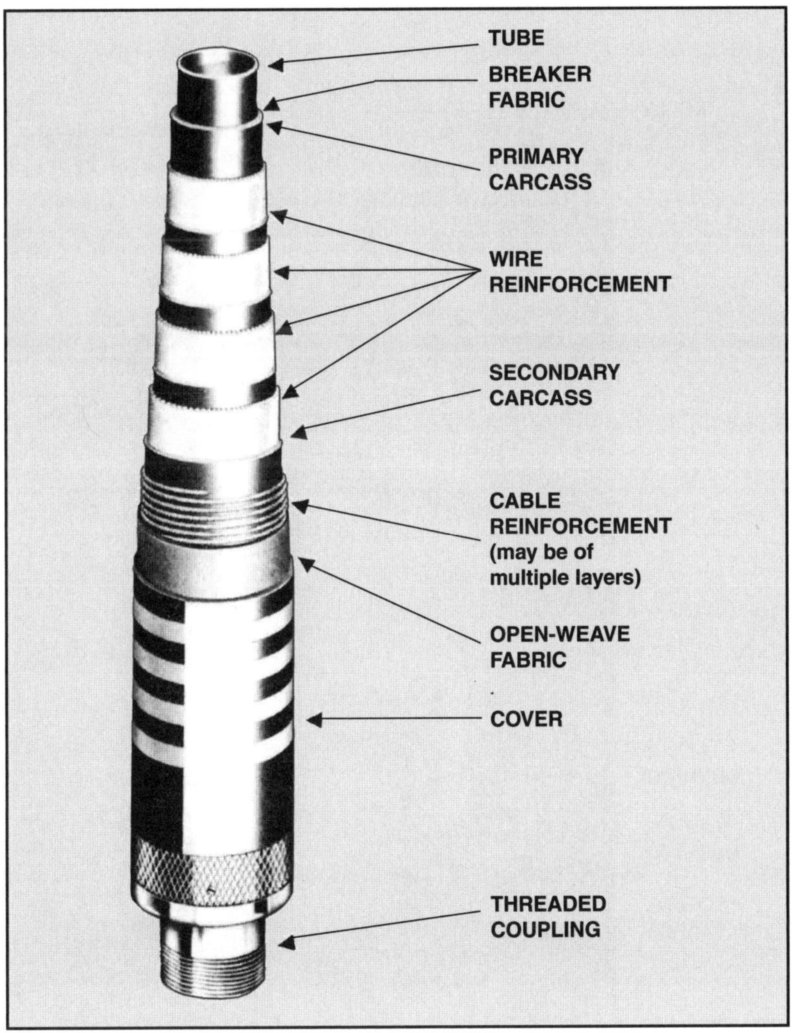

Figure 20. Construction of rotary hose

Rotary hoses generally receive fairly good treatment when in the derrick, but crew members often use poor handling techniques when laying down the hose for rig moves. Proper care includes wrapping the hose on a reel when taking it down and using an air hoist or mobile crane with a cable or sling to keep the hose off the ground. Keeping the hose off the ground prevents damage to the rubber.

Kelly, Drill String, and Bit

The mud then passes through the swivel or the top drive at the end of the rotary hose. If the rig uses a swivel, the next component of the circulation system is the kelly (fig. 21). If the rig uses a top drive, the next step on the circulating path is the drill string. In both cases, the mud flows down the drill string to the bit.

Figure 21. The kelly passes through the kelly bushing, which fits into the master bushing of the rotary table.

After reaching the bottom of the hole, the drilling mud flows out of openings in the bottom of the bit. Depending on the type of bit, the fluid exits either through an outlet in the center of the bit or through jet nozzles that turn the mud stream into high-pressure jets that lift the cuttings off the bottom. The mud also cools the bit. High rotation rates and heavy weight on the bit pressing it into the rock create an enormous amount of friction, which produces heat. So rotation without mud circulation would soon destroy the bit.

Next, the mud with cuttings flows back up the hole through the annulus, the part of the hole that surrounds the drill pipe. At the surface, the mud enters a pipe called the *mud return line.*

Mud Return Line

Mud normally flows out of the return line to the shale shakers. The return line is usually 12 to 14 inches (30 to 35 centimetres) in diameter (fig. 22). A large line is necessary if the mud has to move a long distance from the wellhead to the tanks.

Figure 22. This mud return line is carrying drilling mud from the wellhead, on the right, to a mud-gas separator.

Reserve Pits

Some land rigs with mud tanks also have large, earthen pits bulldozed out of the ground nearby, called *reserve pits*. Waste fluid, cuttings, and even trash that accumulates as a well is drilled go into the reserve pits. Sometimes, the contractor partitions off a small area of a large reserve pit to store surplus good mud not in use at the moment. As a general rule, operators also line the reserve pits with heavy-duty, flexible plastic sheeting to prevent liquid from contaminating the soil. Later, the operator hires a waste disposal company to remove the plastic and haul the waste to a landfill. In places where migratory birds fly over the drill site, the operator may have to place nets over the reserve pits to prevent water birds from landing in them and being harmed.

Layout

In addition to the main parts of the circulating system, a rig has other equipment to treat and route the mud (see fig. 13). These include—

1. hoppers on or near the tanks for adding dry materials to the mud;
2. agitators on the tanks for stirring the mud;
3. auxiliary pumps for sending mud through conditioning equipment for treatment;
4. shale shakers, desander, desilter, mud cleaner, centrifuge, and degasser for treating or conditioning the returning mud;
5. piping and troughs (ditches) to move mud;
6. bins for storing dry bulk materials to be added to the mud;
7. covered storage for sacked materials and other chemical additives; and
8. ample water storage and supply.

Hydraulics of Mud Circulation

Hydraulics is the science that deals with the behavior of a liquid in motion. The hydraulic power of the circulating drilling fluid, in the form of pressure, is very important in drilling—it is what removes the cuttings from the bottom of the hole.

The mud pump is the source of hydraulic pressure for the mud stream. When the mud leaves the pump, it usually has several thousand psi (kilopascals) of pressure, but the pressure drops at different points in the circulation route. The term for this loss of hydraulic pressure is *pressure loss.* Some pressure loss occurs as the mud travels through the surface piping and down the drill stem (fig. 23).

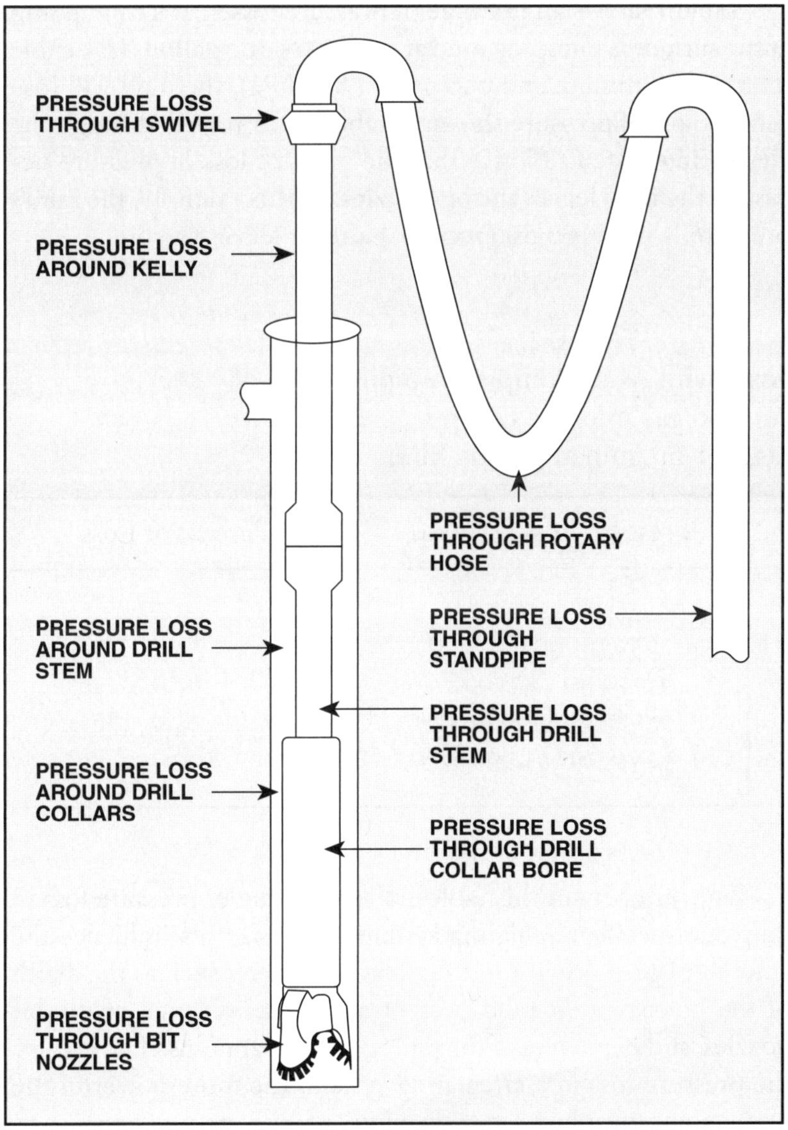

Figure 23. Pressure losses in a circulating mud system

The inside surfaces of these pipes are rough and produce friction and turbulence in the mud stream that reduces pressure. The bit, however, is where the largest pressure loss occurs. Jet nozzles in the bit concentrate the hydraulic pressure so that the mud leaves the bit at a high velocity, or speed, to lift the cuttings off the bottom of the hole. Jet nozzles work like putting your thumb over the end of a garden hose. The water comes out with more force as you make the hole smaller. The remaining pressure after the fluid leaves the bit is just enough to carry the mud up the annulus and back to the surface. Once the mud reaches the surface, all the hydraulic pressure has been lost.

Table 1 shows an example of pressure losses. If a mud pump at the surface is pumping mud at the rate of 400 gallons (1.5 cubic metres) per minute at 2,000 psi (13,800 kPa), the mud will lose the amount of pressure shown in table 1 as it travels through the circulating system. Notice that the greatest loss of pressure occurs as the mud leaves the bit nozzles. In this example, the mud's pressure is only 100 psi (700 kPa) after it leaves the bit.

TABLE 1
Pressure Losses with Mud Pump at Pumping Rate of
400 gal/min at 2,000 psi
(1.512 m^3/min at 13,800 kPa)

Circulation Component	Pressure Loss psi (kPa)	Percent of Loss
Surface equipment	50 psi (350 kPa)	2.5
Drill stem	650 psi (4,480 kPa)	32.5
Bit nozzles	1,200 psi (8,270 kPa)	60.0
Return annulus	100 psi (700 kPa)	5.0
Total loss	2,000 psi (13,800 kPa)	100.0

The numbers in this table are an example of pressure loss in only one specific circulating system. Many factors influence the amount of pressure lost in a circulating system, such as the depth of the hole, type of mud (weight and viscosity), size of the bit nozzles, and diameters of the piping and the annulus. The greater the pressure loss in a circulating system, the more powerful the mud pumps must be to start the mud out with enough pressure to make it back to the surface.

Drilling engineers spend a lot of time studying the factors that affect hydraulic pressure so that they can design a system that makes the most efficient use of the mud pumps' power. If they do their job well, between 50 and 75 percent of the fluid pressure generated by the pump reaches the bit nozzles and cleans the bottom of the hole.

To summarize—

Types of drilling fluids

- Water
- Water mixed with solids and chemicals
- Oil/water emulsions mixed with solids and chemicals

Circulation equipment

- Mud tanks
- Mud pumps
- Standpipe, rotary hose, swivel, drill stem with bit
- Mud return line, reserve pits

Mud treatment equipment

- Shale shaker
- Desander, desilter, mud cleaner, centrifuge
- Mud-gas separator, degasser

Other equipment

- Mud agitators
- Mixing tanks
- Mud hoppers
- Mud monitoring instruments
- Storage facilities, bulk bins
- Water storage and supply

Hydraulics of drilling mud

- The mud pump is the source of hydraulic pressure.
- Hydraulic pressure at the bit nozzles lifts the cuttings.

Air Circulating Systems

Although operators don't use air or gas very often, it is a valuable method of drilling when the formation allows it. Penetration rates are higher because air or gas cleans the bottom of the hole more efficiently than mud. Mud is denser than air or gas and tends to hold the cuttings on the bottom of the hole. As a result, the bit cannot make hole as efficiently because the bit redrills some of the old cuttings instead of being constantly exposed to fresh, undrilled formation.

Air or gas also yields greater footage per bit because drilling mud is very abrasive compared to air or gas and wears out the bit faster. Both air and gas do an excellent job of cooling, and both transport cuttings to the surface quickly. In addition, with air or gas circulation, the formation is easy to identify, and it is easy to detect the presence of gas, oil, or water.

Unfortunately, air or gas drilling has several disadvantages that overshadow the advantages. First, if the formation rock is soft and the walls of the well tend to slough into the hole, air or gas does not have enough hydrostatic pressure to prevent them from doing so. The sloughing walls will probably cause the drill stem to stick. Second, preventing formation fluids from entering the wellbore is impossible because neither air nor gas can exert enough pressure to keep them out. This second disadvantage is especially important because most wells encounter water-bearing formations at some time. Water mixes with the drilled cuttings and balls up the bit, preventing further drilling.

The low hydrostatic pressure also creates the hazard of a blowout if the bit drills into a high-pressure formation. Another disadvantage is that the hazard of fire or explosion is always present. Natural gas is flammable on its own, and air can mix with formation gas and become flammable. Also, corrosion of the drill stem—the reaction of the oxygen in air with metal—can be a problem. Although chemicals to combat corrosion are available, the operator must consider the added cost and effort of using them.

Mist Drilling

When an air or gas drilling operation encounters a formation containing water, the water flows into the wellbore and begins to fill it. This causes two problems. First, because water is heavier than a gas, lifting it out requires more and more gas pressure. Eventually, so much water gets into the hole that the available pressure cannot overcome the weight of the water. Second, the wet cuttings can stick together, fill the annulus, and shut off the flow of air or gas to the surface. The wet cuttings also stick to the bit and drill string. Stuck pipe is the result.

The solution to these problems is to add a foaming agent to the air or gas stream. This is *mist drilling*, or *foam drilling*. The foaming agent is a chemical similar to soap that causes the water in the annulus to froth and foam into a larger volume. Foam is lighter than water because it is full of air, so less pressure can move the water out of the hole. Mist drilling can remove as much as 50 barrels (8 cubic metres) of water per hour entering the hole. Larger amounts of water require switching to aerated mud or ordinary water-base mud.

To add the foaming agent, the crew mixes it with water in a small tank. Then a small pump pumps the agent into the air stream going into the well (fig. 24), along with corrosion-inhibiting chemicals. The foam, like plain air, flows away at the surface through a blooey line.

Figure 24. Chemical tank and pump for circulating with foam

Equipment

As mentioned earlier, a gaseous drilling fluid does not actually circulate. Instead, it follows the same path down and back up the hole as drilling mud but then flows out through a blooey line. At the end of the blooey line a pilot light ignites the gas as it leaves. This is flaring the gas. Operators usually set a flare even when using air or mist because some natural gas may have entered the air during drilling. Crew members extend the end of the blooey line a good distance from the rig to reduce the fire hazard and to keep the cuttings from blowing onto the rig floor.

The second main difference between air and mud circulation is that pumps add pressure to the mud, whereas a *compressor* adds pressure to the air. As the name indicates, a compressor compresses, or adds pressure to, the air. Usually, operators use compressors only when air is the circulating fluid. Normally, they use natural gas only if a high-pressure source is available nearby, and in that case no further compression is necessary. A regular rotary rig arranged for *air drilling* uses compressors mounted on *skids*, or portable platforms.

Other equipment that the contractor may use to handle air or gas for circulation includes chemical treatment equipment to combat corrosion, mist pumps for injecting foaming agents and fluid, and air bits. Air bits are extra-heavy bits that allow air to cool the bit bearings.

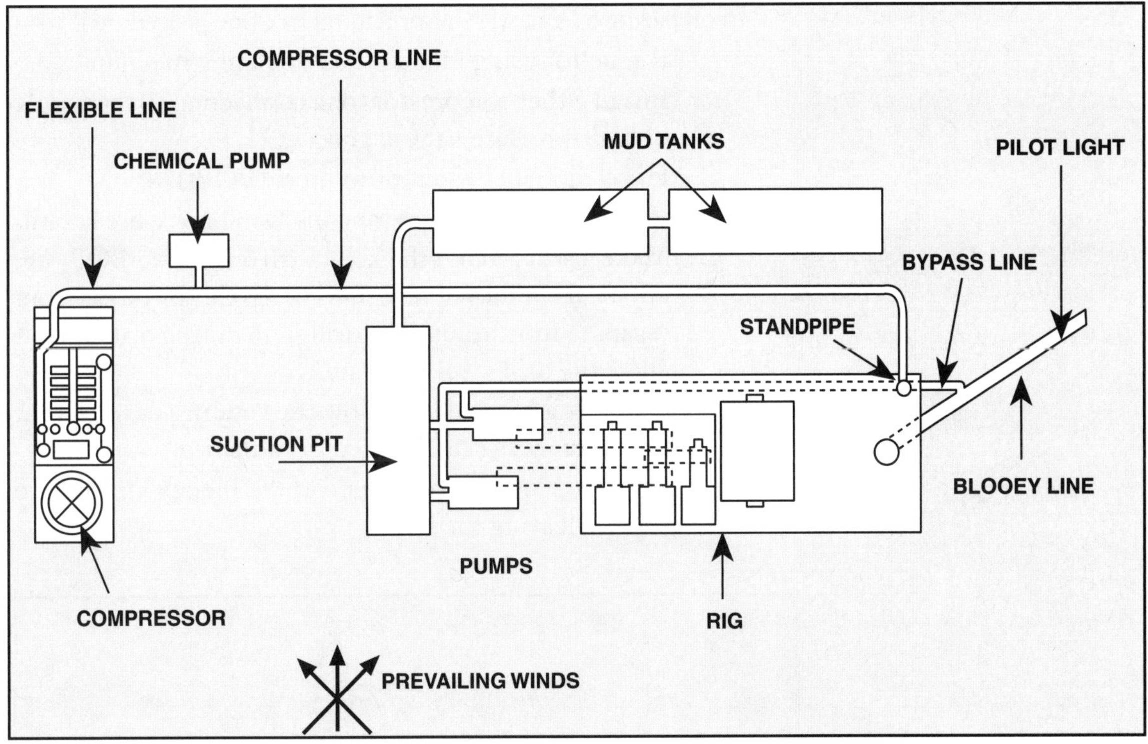

Figure 25. Arrangement of equipment for air circulation

Layout and Drilling Procedures

Fig. 25 shows a typical layout for air circulation equipment. Notice that the line from the compressor bypasses the mud tanks and pumps and connects directly to the standpipe. A bypass line from the wellhead carries the air returning up the annulus to the blooey line.

Because natural gas is flammable, drilling with gas requires special safety precautions. The same precautions are necessary when drilling with air because natural gas from the formation may enter the circulating air. Never smoke on a rig drilling with air or gas. All electrical equipment should be explosion-proof to prevent sparks.

In preparing to drill with air or natural gas, one procedure is to—

1. Rig up for using mud as the circulating medium.
2. Install either compressors or a connection to the supply of high-pressure natural gas.
3. Hook up the blowout preventers (BOPs).
4. Put into effect safety precautions to minimize fire hazard.
5. Make a seal around the kelly with a rotating BOP (fig. 26). Seals in the rotating BOP keep the air or gas from escaping from around the drill stem during drilling, but allow the drill stem to rotate.
6. Connect a blooey line below the rotating head to vent the air or gas a safe distance from the rig.
7. Set up a pilot light at the end of the blooey line to flare any gas leaving the hole.

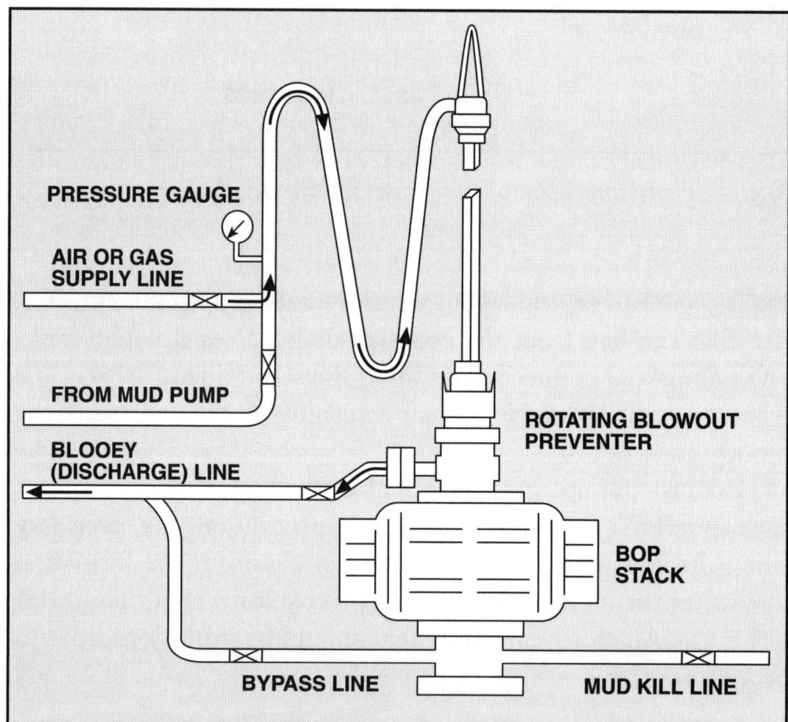

Figure 26. Blowout preventers for air drilling

Volume, Pressure, and Velocity

Volume, pressure, and velocity are the variables in air or gas drilling, and they are all related to each other. Volume and pressure directly relate in the following way. Imagine a balloon filled with air. If you squeeze the balloon and make it smaller, the pressure inside increases. Squeezing it enough eventually causes the pressure to overcome the strength of the rubber—the balloon bursts. Reducing the volume raises the pressure. Just as squeezing a balloon raises the pressure inside the balloon, compressor pistons moving back and forth inside cylinders raise the pressure of the air the compressor takes in. The compressor cylinders confine the air. The pistons move to decrease the amount of space the air takes up in the cylinders and raises the pressure.

To lift cuttings, the pressure and velocity of the air or gas must be high. In a gas at room temperature and pressure, the molecules are much farther apart than the molecules in a liquid. The compressor squeezes the gas's molecules closer together, which adds energy in the form of velocity. Natural gas must be even more compressed (and therefore the volume must be greater) than air to reach the same velocity, because it is lighter than air.

The velocity of the air or gas returning up the annulus depends on several variables. The most important are the rate of penetration, well depth, and the amount of water entering the well. Other variables are the diameters of the hole and drill pipe, type of formation being drilled, and size of cuttings. As you can see in table 2, the deeper the well and the greater the penetration rate, the more air or gas is needed to keep the velocity high enough to lift the cuttings. All depths and penetration rates require more gas than air.

TABLE 2
Approximate Rate of Circulation (ft³/min and m³/min) in 8¾-inch hole (m³/min in 222-mm hole) with 4½-inch (114-mm) Drill Pipe with Volumes to Produce Lifting Power Equivalent to a Velocity of 3,000 ft/min (915 m/min)

ROP	30 ft/min (9 m³/min)	90 ft/min (27 m³/min)	30 ft/min (9 m³/min)	90 ft/min (27 m³/min)
Well Depth	Compressed Air		Natural Gas	
	Circulation Rates			
2,000 ft (609.6 m)	1,039 ft³/min (29.421 m³/min)	1,113 ft³/min (31.517 m³/min)	1,326 ft³/min (37.548 m³/min)	1,426 ft³/min (40.380 m³/min)
4,000 ft (1,219.2 m)	1,174 ft³/min (33.244 m³/min)	1,323 ft³/min (37.463 m³/min)	1,486 ft³/min (42.079 m³/min)	1,648 ft³/min (46.666 m³/min)
6,000 ft (1,828.8 m)	1,310 ft³/min (37.095 m³/min)	1,573 ft³/min (44.542 m³/min)	1,646 ft³/min (46.610 m³/min)	1,946 ft³/min (55.105 m³/min)

Compressors

Compressors may be single-stage, two-stage, or three-stage. Each stage compresses the air more than the last. They can move air at 50 to 3,000 cubic feet per minute (1.4 to 84 cubic metres per minute) with pressures up to 3,500 psi (24,133 kPa). If the rig is circulating foam or aerated mud, or if the hole encounters water, the contractor will hook up booster compressors. Booster compressors take the output of one or two units and raise the pressure.

To summarize—

Types of drilling fluids

- Air
- Natural gas
- Mist or foam
- Air mixed with water

Circulation equipment

- Compressors
- Standpipe, rotary hose, swivel (or top drive), drill stem and bit
- Blooey line

Volume, pressure, and velocity

- Reducing the volume of a gas (compressing it) raises its pressure.
- A high volume of pressurized gas has velocity for circulation and lifting cuttings.

Drilling Muds

The type of drilling fluid the operator uses on a particular well depends on many factors. Some of them are the formations being drilled, whether the hole encounters water, the pressure in the hole, and how deep the hole is. One overriding concern is the rate of penetration. Generally, the faster the penetration rate, the less expensive it is to drill the hole. So operators tend to use the lightest possible fluid that can control the conditions downhole. Air or gas provides the highest ROP (rate of penetration). Unfortunately, operators cannot use air or gas in too many holes because of the limitations air or gas impose on drilling.

The most common drilling fluid is drilling mud, which is a mixture of water or oil with additives that help mud do its job better. Crew members usually have to alter the proportion of additives as the hole deepens and encounters new formations. So the derrickhand or a mud specialist from the company that supplied the fluids and additives, the *mud man*, *mud engineer*, or *mud hound*, tests the mud at regular intervals. The mud specialist adjusts the amount and composition of chemical additives as required.

Composition of Drilling Muds

Drilling mud is a three-phase mixture. Three-phase means that it contains three types of material. Drilling muds have a liquid phase and two solid phases. The liquid phase, or *continuous phase*, may be either water or oil or a water-oil mixture. One of the solid phases contains solids that can react with the liquid. This is the *reactive*, or *colloidal*, *phase*. The main reactive solids in most drilling muds are clays. Clays react with water by swelling to thicken the mud.

The other solid phase is made up of solids that do not react with the liquid, the *nonreactive phase*. The nonreactive solids in drilling mud include finely ground cuttings, or drilled solids, such as sand, chert, limestone, dolomite, some shales, and mixtures of many minerals. These solids are undesirable because they are abrasive and can damage pump parts. Also, they add extra weight to the mud. Extra weight can slow down the penetration rate of the bit. The mud may also contain certain nonreactive solids that crew members purposely add to alter its properties. One common nonreactive solid is *barite*, which increases mud weight to control formation pressures.

Solids in Mud

Mud solids come in many different sizes and densities (specific gravities). Because solids have different sizes and gravities, manufacturers can build mud conditioning equipment to separate desirable from undesirable solids. The proportion and types of solids in drilling mud greatly affect its properties. Because mud engineers must control these properties, they install equipment to remove the drilled solids and use special techniques to control the amount and types of added solids.

Size

API classifies the solids in mud by their size. Often, technicians use screens with different sized openings to size solids. They speak of a screen's fineness or coarseness in terms of *mesh*, or how many open-ings per square inch the screen has. For example, a 100-mesh screen has 100 openings per square inch. It has a larger mesh (it is coarser) than, say, a 400-mesh screen, which has 400 openings per square inch.

API classifies any particle that will not pass through a 200-mesh screen as *sand*, regardless of what it is made of. So "sand" in this case may be particles of any sort of rock as well as actual sand. Sand is larger than 74 microns and smaller than 2 millimetres. As mentioned earlier, micron is short for micrometre, which is one millionth of a metre, or 0.000039 inch. Sand feels gritty between the fingers. Particles smaller than sand are called *silt*. Silt is so small it feels soft between the fingers. However, any drilled solids larger than 10 to 15 microns can be abrasive, so both sand and silt can damage equipment. The very smallest particles are *colloids*, smaller than 2 microns. They are invisible as individual particles. Because of their size, colloids have special properties: they do not settle out of the mud by gravity and they are small enough to pass through filter membranes.

Specific Gravity

Specific gravity is a way engineers define the density of a solid or liquid. The specific gravity of a fluid is a ratio of the density of a fluid to that of a reference fluid. For example, pure distilled water is the reference fluid for liquids and solids. Water has a specific gravity of 1. Thus, any substance (liquified or solid) with a specific gravity of less than 1 is lighter (less dense) than water. Conversely, any substance with a specific gravity more than 1 is heavier (denser) than water. Barite, for example, has a specific gravity of 4.2; it is over four times denser than water. Solids in mud are either low-gravity or high-gravity; low-gravity solids are lighter than high-gravity solids. Drilled solids and clays are low-gravity solids. One high-gravity solid is barite, a material for weighting mud.

Water-Base Muds

Operators use *water-base muds* more than any other drilling fluid. Because either fresh water or salt water is available in most of the world's drilling areas, water-base muds are easier to use and less expensive than oil-base muds or compressed air. The main types are natural muds, treated muds, spud muds, saltwater muds, and aerated muds.

Natural Muds

Sometimes formations near the surface contain enough clay to make up a good *natural mud* when mixed with water. Operators often use natural mud both for surface drilling and for making hole rapidly below the conductor casing. (Conductor casing is the first short string of casing placed in the wellbore.) The crew usually adds chemicals to natural mud to improve its properties as the depth of the hole increases. Operators have drilled wells to 12,000 feet (3,658 metres) on the U.S. Gulf Coast using natural mud. Crew members added a little commercial bentonite (a clay) to thicken the mud and give it the capability of laying down a filter cake on the wall of the hole. As mentioned earlier, this wall cake stabilizes the wall of the hole and keeps it from caving in.

Treated Muds

Crew members add chemicals and minerals to most water-base drilling muds to change its properties and its reaction to the formation. Table 3 lists many additives crew members use to affect different properties. Mud chemistry is complex, as you will see, so always follow the mud engineer's instructions carefully when adding and mixing additives.

TABLE 3
Basic Mud Chemicals

Viscosifiers
- Bentonite
- Attapulgite
- Polymers

Viscosity-Reducing Chemicals
- Lignosulfonate
- Lignites
- Tannates
- Phosphates

Fluid-Loss Reducers
- Starches
- CMC
- Bentonite
- Synthetic polymers
- Lignites
- Lignosulfonate

Weighting Materials
- Barite
- Galena
- Calcium carbonate
- Dissolved salts
- Iron oxide

Swelling Inhibitors
- Salt
- Encapsulating agent
- Lime
- Gypsum

Emulsifiers
- Lignites
- Lignosulfonate
- Detergents

Lost-Circulation Materials
- Granular
- Fibrous
- Flaked
- Slurries

Special Additives
- Flocculants
- Corrosion control
- Defoamer
- pH control
- Mud lubricant
- Antidifferential sticking material

Increasing Viscosity. If the surface formation cannot produce a good natural mud, crew members may add bentonite or another reactive clay to fresh water. Reactive clays swell when wet, forming a gel (fig. 27). Clay increases the viscosity of the mud and builds filter cake to decrease fluid loss. It is an important component of most drilling muds. Other additives that increase viscosity include attapulgite and various polymers.

Figure 27. Bentonite reacting with water

Attapulgite is a special clay crew members add to saltwater muds to increase their viscosity and wall building properties. Operators do not, however, use it as much as they used to. Mud companies have discovered better chemicals (polymers) to build viscosity in saltwater muds. Operators can also use polymers in freshwater muds.

Polymers are a class of chemicals that may be synthetic or natural. A polymer is a substance that consists of large molecules formed from smaller molecules in repeating structural units. Polymers are like a long chain of molecules strung together. Polymer additives to mud include *carboxymethyl cellulose (CMC)* and starches. They form even more powerful colloids and gels than bentonite and dissolve in water. In this way, the mud's viscosity increases without added solids, so the operator may decide to add less bentonite along with a polymer. Polymers affect other properties of the mud besides viscosity, so they serve more than one function.

Decreasing Viscosity. The viscosity of a mud can also be too high. Natural muds or muds containing contaminants such as cement or drilled solids may be too viscous. Also, a sticky formation geologists call *gumbo* can react with the mud and thicken it. Mud that is too thick, or viscous, is more difficult for the pump to move and has high pressure losses. If pressure losses are too high in the drill stem, there may not be enough pressure left at the bit nozzles to adequately clean the bottom of the hole.

To thin a water-base mud, the crew can add either more water or chemicals called *dispersants*. Adding water lowers the mud's viscosity but also reduces its density. After thinning the mud with water, crew members must add weighting material to raise the density back to the desired figure.

Mud companies market special chemical thinners (dispersants) that reduce viscosity without reducing density. When using weighted muds, the crew may therefore add a dispersant rather than water. Generally, when the percentage of solids is in the correct range, they usually add a dispersant. If the percentage of solids is high, they usually add water.

Lignosulfonates, the most widely used dispersant, are very effective but are also expensive. Operators can use them in all water-base mud systems. Other dispersants include lignites, tannates, and phosphates. The lignites are the next most popular after lignosulfonates; phosphates and tannates are now rare.

Some dispersants do other jobs as well, which may be more important than reducing viscosity. Both lignites and lignosulfonates reduce fluid loss and thus the thickness of the filter cake, counteract the effects of salts, minimize the effect of water on the formation, emulsify oil in water, and keep the mud from breaking down (stabilize the mud) at high temperatures.

Controlling Fluid Loss. Lower fluid loss prevents the mud from getting too thick because of decreased water content. Additives that control fluid loss pack tightly together to form a good filter cake. Materials that control fluid loss include bentonite, lignites, lignosulfonates, CMC, synthetic polymers, and starches. Besides controlling fluid losses, these materials have other uses as well, such as reducing viscosity and increasing gel strength.

Controlling Mud Weight. Among the first materials drillers used to increase the weight, or density, of drilling fluids was iron oxide. The most popular iron oxide they used was hematite. Today, well owners mostly use barite, or naturally occurring barium sulfate as a weighting material in all types of treated mud. Barite has low cost, high specific gravity, is inert (nonreactive), and occurs naturally in a pure state. Barite is a fine grayish-white powder that is 4.2 times denser than water. Other weighting materials operators use include galena (a lead ore), calcium carbonate, and various salts, such as sodium chloride, calcium chloride, and the like.

In most of the U.S., personnel measure mud weight in pounds per gallon (ppg). On the Pacific Coast of the U.S., they often use pounds per cubic foot (pcf). Practically everywhere else, crew members use either kilograms per cubic metre (kg/m^3) or ppg. Normally, mud weights range from about 8.33 ppg (62.3 pcf or 998 kg/m^3) to 20 ppg (149.6 pcf or 2,397 kg/m^3). A mud weight of 16 to 18 ppg (119.7 to 134.7 pcf or 1,917 to 2,157 kg/m^3) is considered heavy.

Inhibiting Swelling of Water-Sensitive Formations. Water-base drilling fluids can cause formations that are sensitive to water, such as shales and clays, to swell and slough into the hole (fig. 28). Swelling shales and clays can impede drilling, enlarge the hole, increase viscosity, and damage oil-producing zones. Adding lime (calcium hydroxide), gypsum (calcium sulfate), calcium chloride, or some other calcium compound to inhibit this swelling was common until the 1970s.

Operators still use calcium muds, also called *lime muds* or *gyp muds*, but today they mostly use an oil mud or a nondispersed water-base mud containing a salt and an *encapsulating agent*. An encapsulating agent is a chemical that surrounds (encapsulates) drilled solids. Encapsulating them makes it easy to circulate them to the surface where the solids removal equipment can get them out of the mud. An encapsulating agent also coats water-sensitive shales to cut back on (to inhibit) their swelling from contact with water. One mud that encapsulates drilled solids and inhibits swelling is a mud to which crew members have added potassium chloride (KCl) and a polymer (often, polyacrylamide) for encapsulation. Personnel usually refer to such muds as KCl/polymer muds.

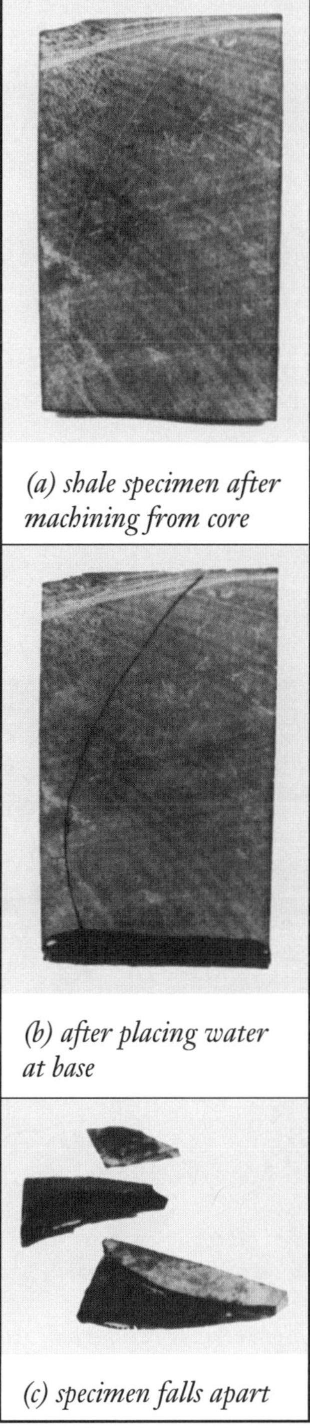

(a) shale specimen after machining from core

(b) after placing water at base

(c) specimen falls apart

Figure 28. Destabilization of shale specimen by hydration of a healed fracture (Courtesy of Gulf Publishing Company)

Lubricating. An *oil-emulsion mud* is a water-base mud with a little oil mixed in, usually a low-toxicity oil (fig. 29). Oil-emulsion muds increase drilling rate, reduce *torque* (twisting force) on the rotating drill string, reduce sticking of the drill pipe, and increase the life of the bit because the oil has a lubricating effect.

When crew members add oil to a water-base mud, they usually add an emulsifier. Adding an emulsifer creates an emulsion of oil and water. An *emulsion* is a stable mixture of two liquids that normally do not mix, like oil and water. Normally, oil and water separate as soon as the movement of circulating stops. The addition of an emulsifier keeps them mixed.

Normally, crew members can add as much as 5 percent oil to a water-base drilling mud without an emulsifier. However, adding an emulsifier to a mud with even low amounts of oil stabilizes the mixture—ensures that the oil does not drop out of the water-base mud. Lignites, lignosulfonate, or soap-type additives are good emulsifying agents.

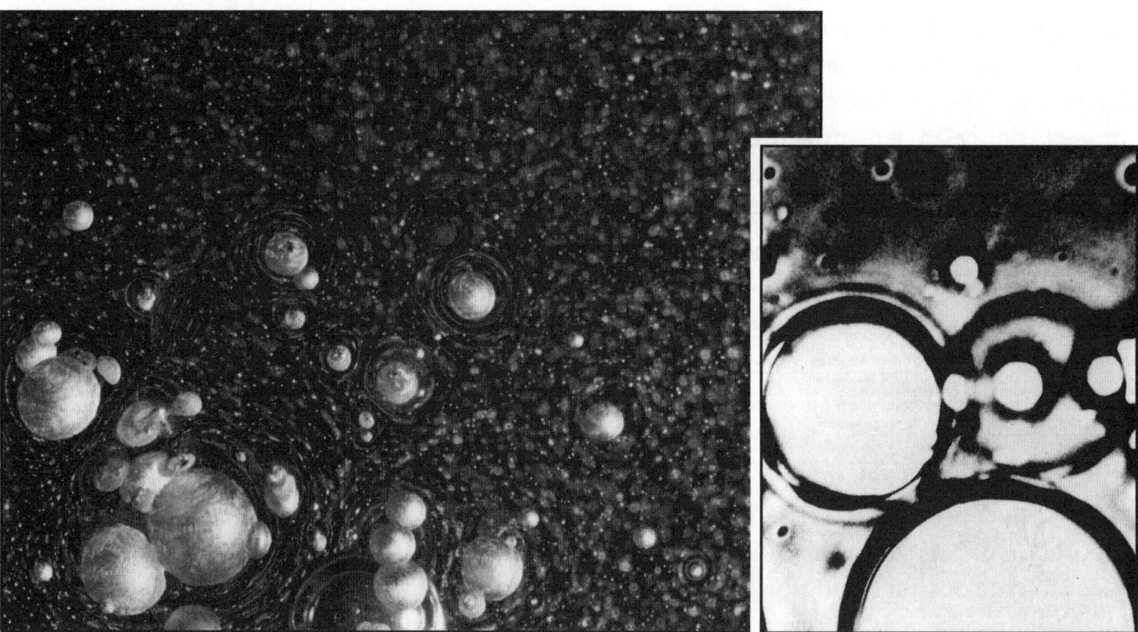

Figure 29. Two percent oil emulsion mud magnified 900 times

Preventing Lost Circulation. Lost circulation, or lost returns, happens when quantities of whole mud—the liquid part as well as the solid parts of the mud—flow into a formation. Do not confuse lost circulation with fluid loss. Fluid loss occurs when hydrostatic pressure in the hole forces the liquid part of the mud, the filtrate, into an adjacent porous and permeable formation. A certain amount of fluid loss is desirable because it leaves a thin, impermeable filter cake on the wall of the hole. The filter cake stabilizes the hole and prevents further fluid loss when it reaches a certain thickness.

Lost circulation can occur in fractured formations, where geologic forces created natural openings. Sometimes the fractures are large, sometimes small, and sometimes in between. Regardless of the size of the fractures, mud flows into them. If the fractures are large, lost circulation may be complete; no mud returns to the surface. At other times, lost circulation is partial; some mud returns to the surface.

A mud weight that is too high can also cause lost circulation. In this case, the hydrostatic pressure of the mud in the hole is higher than the formation pressure. If this pressure differential is high enough, the pressure fractures the formation and whole mud flows into the formation. The fracturing may be light or severe. Whether light or severe depends on how high the pressure differential between the mud and the formation is and how sensitive the formation is to being fractured by differential pressure. High differential pressure on an easily fractured formation can mean trouble—that is, complete loss of circulation.

Where differential pressure causes lost circulation, crew members can keep it from happening by reducing the weight of the mud and thus its hydrostatic pressure. Where natural fractures cause lost circulation, the crew can often stop lost returns by adding a material to the mud that plugs up (seals) the fractures. *Lost circulation materials (LCM)* come in a variety of types and sizes—grains, flakes, or fibers, and coarse, medium, or fine. Granular materials include ground-up walnut or pecan shells. Polyester or plastic flakes come in sizes from ⅛ inch to 1 inch (3 to 25 millimetres) long. Fibrous materials may be wood or plant fibers or synthetic fibers. Most commonly, crew members add a mixture of types and sizes of LCM to cure lost circulation.

Occasionally, the lost circulation rate is so high that the crew cannot seal the fractures. If the formation does not contain fluids under pressure, they may be able to drill blind. *Blind drilling* is drilling without the drilling mud returning to the surface. When drilling blind, the driller pumps fluid (usually water) down the drill stem and into the formation. The fluid moves the cuttings away from the bit but does not return them to the surface.

pH Control. The *pH* of a fluid is a measure of its acidity or *alkalinity*. Technicians measure the pH of a liquid on a number scale that ranges from 0 to 14. A neutral liquid, such as pure water, has a pH of 7. An acid solution has a pH less than 7; an alkaline solution has a pH more than 7. Usually, mud must be very alkaline to allow the chemicals in the mud to work well and to minimize corrosion. For example, lignosulphonates and lignites need a pH of at least 9 to work as dispersants (thinners).

A drilling fluid with a pH below 7 is acidic. Acids speed up the chemical reaction that corrodes steel. Alkalines, on the other hand, slow down corrosion. An acidic mud would accelerate corrosion of the drill pipe and other steel tools in the hole. But an alkaline mud, (one with a pH of 10 to 12) minimizes corrosion. Operators there-fore make the drilling mud alkaline.

Many additives change the alkalinity of the mud. Mud containing bentonite usually has a pH of 8.5 to 9.5. Lime systems have a pH of around 12. But sometimes the crew must add a *pH control agent* to change pH. The most commonly used pH control agent is caustic soda (sodium hydroxide). Caustic soda is extremely alkaline and can cause severe burns if it contacts the skin. Crew members therefore handle it with care. They wear proper *personal protective equipment (PPE)*, such as respirators, face shields, and full-body protection to keep caustic soda off their skin.

Other Additives. The crew may add a number of other substances to improve the properties of the drilling mud. *Flocculants* cause drilled solids to stick together and clump into larger pieces that are easier to remove than very small particles. *Corrosion-control agents* reduce corrosion of the metal parts that salts in the mud can cause.

Defoamers break up foams that form when formation gas contaminates the mud. *Lubricants* reduce friction, which helps the bit to wear longer and the drill string to turn more easily against the wall of the hole. Finally, *antidifferential sticking additives* help pre-vent the drill string from getting stuck against the side of the hole. Crew members can also add such additives to help free pipe that is already stuck. *Differential sticking* (wall sticking) can occur when the drill stem, usually the drill collars since they are large in diameter, contacts the wall of the hole where it exposes a porous and perme-able formation. When the hydrostatic pressure of the mud in the hole is higher than formation pressure and when the drilling mud's fluid loss is high, the higher hydrostatic pressure causes the drill stem to stick. Differential pressure adheres the drill stem to the side of the hole.

Spud Muds

The composition of the fluid operators use to start a well, the *spud mud*, varies with drilling practices around the world. Sometimes the operator uses water alone from a nearby source such as a well, stream, or lake. In some places, however, surface formations may consist of loose sand and gravel. In such cases, the spudding requires a fairly good quality of mud. This spud mud must be able to build up a good filter cake on the wall of the hole to prevent caving. It must also be viscous enough to carry out cuttings as the hole is drilled. Bentonite is a common additive to spud muds to control viscosity.

Sometimes operators add natural gums into the water making up the spud mud. One is guar gum. Its main advantage is that it produces high viscosity spud muds in either fresh or salt water. After drilling past the conductor pipe, the operator usually discards or recycles the spud mud because cement often contaminates it. The mud is therefore not suitable for deeper formations. Cement contamination comes from the cement crew members use to cement the conductor pipe or the surface casing to the wall of the conductor or surface hole. During the cementing process, cement forces the spud mud out of the hole as it replaces the mud around the casing. The cement mixes with the mud where the two fluids come into contact and contaminates the mud.

Saltwater Muds

When the available water is salt water, such as in offshore wells, operators use it. Salt water is an excellent source of calcium and magnesium, which controls viscosity. Operators often spud offshore wells with seawater mixed with polymers or guar gum. Later they improve the mud with bentonite, caustic soda, and lignite or lignosulfonate. Sometimes they add gypsum or an inhibitive salt for stability. Starches control fluid loss.

On land, operators may drill with a saturated saltwater mud. One use of saturated salt water is to drill a bed of salt. If the operator used a freshwater mud to drill a salt bed, the hole would enlarge because the fresh water in the mud dissolves the salt. Saltwater mud already has as much salt dissolved in it as it can hold, so it overcomes this problem.

Operators sometimes use saturated or nearly saturated salt water for completing the well. That is, when the borehole is about to enter the producing formation, the operator may use a saltwater mud to prevent formation damage. If the formation contains water-sensitive shale, for example, fresh water causes it to swell. The swollen shale then blocks the formation's permeability. Blocked permeability can reduce or totally prevent the flow of oil and other fluids into the well. To prevent such damage, operators may use salt water because salt water does not cause shale to swell.

Aerated Muds

Another option is *aerated mud*, a fluid that forms when both air and mud are pumped into the standpipe at the same time. It looks like gas-cut mud. *Gas-cut mud* is drilling mud that contains *entrained* formation gas (tiny gas bubbles). This entrained gas often gives the mud a fluffy or bubbly texture. Aerated mud may replace regular drilling mud to prevent lost circulation and increase the penetration rate. As mentioned earlier, one of the causes of lost circulation is drilling mud whose hydrostatic pressure is so high that it fractures the formation. Air or gas in the mud lightens it and so reduces the amount of pressure the mud exerts on the formation. When drilling with air or gas, the operator may switch to aerated mud if more than 50 barrels (8 cubic metres) per hour of water enter the hole.

Oil Muds

Oil muds have oil, usually diesel fuel or a low-toxicity oil, as the liquid phase instead of water. A low-toxicity oil is a lightweight oil, such as diesel fuel, from which the refiner has removed the harmful components. Some water is always present, so all oil muds must contain an emulsifier.

Oil muds are much more expensive and harder to handle than water-base muds, so operators use them only when downhole conditions require it. When water-base muds would cause a problem—in formations containing water-sensitive shales, corrosive gases, or water-soluble salts, for example—an oil mud can be the answer. Oil muds also have the advantages of a higher boiling point and better lubricating ability. So, they are useful in high-temperature wells. They also minimize stuck pipe as well as torque and drag in directional drilling.

The disadvantages of an oil mud are its expense, the problem of contamination by groundwater (increased water content causes the mud to thicken), its flammability, and, usually, slower drilling rates. In addition, the operator must take precautions to prevent the mud from contaminating the environment during use, and dispose of it in a safe manner.

The two types of oil muds are oil-base and invert-oil muds.

Oil-Base Muds

An *oil-base mud* consists of oil, emulsifiers, stabilizing agents, salt, and less than 5 percent water. Although oil-base mud has a small amount of water, too much water contaminates the mud. If additional water gets into the mud and raises the water content to 5 percent or higher, the mud gets too thick to pump efficiently.

Invert-Oil Muds

To overcome the problem of water contamination, researchers added emulsifiers to oil muds. Emulsifying the water in the oil made the water a useful component instead of a contaminant. In short, they made an oil-emulsion mud. An oil-emulsion mud is an oil mud in which the water is spread out (dispersed) in the oil. The water occurs as small droplets suspended in the oil. Mud engineers coined the term *invert-oil muds* for oil-emulsion muds. Invert-oil muds contain from 10 to 60 percent water, all of which is emulsified in the oil. Some of the advantages of an invert-oil mud are that any mud with more than 30 percent water will not burn and it costs less than oil-base muds.

An invert-oil mud is composed of oil, water, emulsifiers, and stabilizing agents. The viscosity is decreased with oil and increased with water or organophilic clays. (An organophilic clay is a treated clay that is attracted to oil rather than water.) A disadvantage of invert-oil muds is that they have a high fluid loss, but emulsifiers, stabilizing agents, and other special compounds can control it.

A relaxed invert-oil mud generally has a higher oil-to-water ratio than a regular invert-oil mud. Otherwise its composition is the same as an invert-oil mud. Advantages include a penetration rate as good as or better than a water-base mud, resistance to contaminants, and stability at high temperatures.

Additives to Oil Muds

The crew usually mixes oil muds on site. Both oil-base muds and invert-oil muds require the addition of a wetting agent, a viscosity increaser, lime, and a weighting material. Other additives work as viscosity reducers and fluid-loss control agents.

Wetting Agents. Most minerals used as mud additives, such as barite, clays, and lime, are *hydrophilic* (literally, water-loving). Hydrophilic additives thus attract any bit of unemulsified water in the mud, which causes the additives to clump and settle out. Hydrophilic additives also make the mud's viscosity and gel strength too high, gum up screens that filter solids out of the mud, coat the drill pipe, and destabilize the emulsion. To combat these effects, crew members add a *wetting agent*, or *surfactant*, to oil muds. A surfactant, which is short for surface active agent, is a chemical that adheres to the surface of a substance. In an oil mud, the surfactant bonds to the hydrophilic solids in the mud (fig. 30).

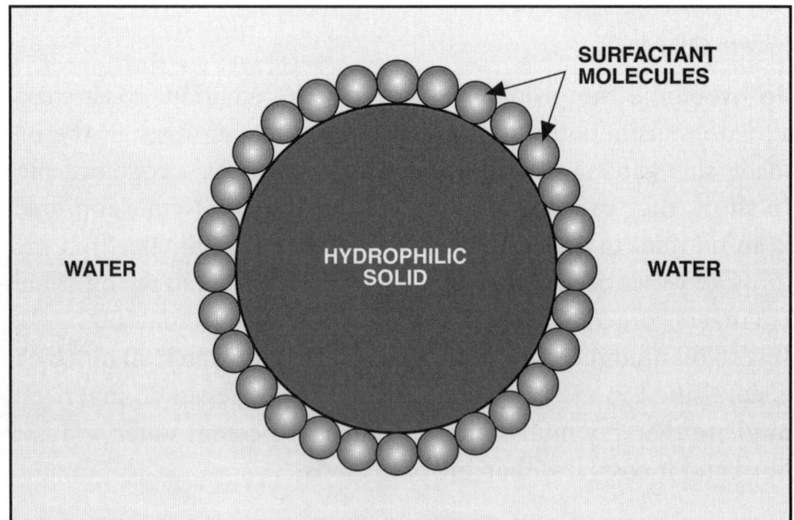

Figure 30. Protective skin of surfactant molecules around a hydrophilic solid

The surfactant keeps water away from the solids so that they do not clump together.

Organophilic Clays. For controlling viscosity, oil muds cannot use ordinary bentonite because bentonite is hydrophilic. That is, it is attached to any water in the oil mud. The water causes the bentonite to clump together and form a thick mass.

To produce a clay for viscosity control in oil muds, the supplier adds an *amine salt* to ordinary bentonite or other clay. Amine salt is an organic compound derived from ammonia, in which organic compounds replace one or more of the hydrogen atoms in the ammonia. Amine salt converts a hydrophilic clay to an *organophilic clay*. An organophilic clay is a clay that is attracted to organic substances such as oil. Since it is attracted to oil, instead of water, the organophilic bentonite does not clump up and settle out. Suppliers sell these treated clays under the label of organophilic clays, organic clays, or amine clays.

Lime. Lime (calcium hydroxide) in oil muds is an emulsifier. It also combats contamination by carbon dioxide (CO_2) in the air and hydrogen sulfide (H_2S) in petroleum. Lime reacts with these gases to produce water and a harmless salt. Oil muds usually contain lime as a preventive measure, just in case the borehole encounters H_2S.

Weighting Materials. Crew members usually add barite to weight up oil muds, just as they do for water-base muds. Instead of barite, they may use hematite. It is not only a good weighting material, but also it maintains a lower solids content. A few oil muds contain calcium carbonate as the weighting material, which helps prevent lost circulation as well.

Other Additives. Crew members may add ammonium humate or asphaltenes for fluid-loss control. Asphaltenes form long, sticky molecules that seal extremely small openings in the formations, bringing fluid loss to near zero.

To reduce viscosity, the crew may add an organic sulfonate. However, relying on such a remedy is tricky. Organic sulfonates work by stripping the amines from treated organophilic clays, so the crew could ruin the mud by adding too much sulfonate and then compensating by adding more clay, and so on. Also, if finely drilled solids caused the increase in viscosity, adding sulfonates can increase the viscosity even further.

Synthetic Muds

Operators, rig owners, and rig personnel need to be aware of the effects of drilling mud on the environment. Almost all drilling muds contain components that are toxic to human, animal, and plant life. Oil muds are especially damaging. Some of the harmful effects are invisible and some are observable. Unseen effects of mud, for example, include pollution of groundwater, which may be a drinking water supply. Offshore, release of mud can damage marine life. Visible effects include pollution of surface water and land, affecting soil productivity. The U.S. and other governments have enacted legislation to protect the environment, and rig personnel must be familiar with the laws in order to comply.

At the same time, the search for petroleum has resulted in drilling deeper wells in increasingly difficult environments, such as in deep water and arctic locations. These locations present special problems—for example, high downhole pressures and temperatures. Traditional oil muds that can handle these adverse conditions are also the most environmentally hazardous.

So engineers developed synthetic muds that have a lower impact on the environment and are easier to dispose of safely. For example, some new invert-oil muds use vegetable oil or ester oil, both of which are biodegradable. Others use mineral oil, which is less toxic than diesel oil.

Mud companies have also developed environmentally safe additives for water-base muds. Such additives include PAL (polyanionic lignin), polymers such as VA/VS (vinyl amide/vinyl sulfonate copolymer) and MPT (modified polyacrylate terpolymer), and an oxygen scavenger to control corrosion. (Corrosion can be a big problem at high temperatures.) Using environmentally safe muds lowers the normally high cost of disposing of cuttings, which are coated with the mud. Also, programmers have developed computer software that helps operators understand and monitor compliance with environmental regulations.

To summarize—

Main components of drilling muds

- Continuous phase
  - Fresh or salt water
  - Oil
  - Mixture of fresh water and oil
- Reactive or colloidal phase
  - Clay (usually bentonite)
  - Dispersant, flocculant, corrosion control agent, pH control agent, defoamer (water-base muds)
  - Emulsifier, surfactant (oil muds)
- Nonreactive phase
  - Drilled solids (sand and silt)
  - Weighting material (usually barite)
  - Material to prevent lost circulation

Testing of Drilling Mud

On land rigs, derrickhands (and offshore, on some rigs, *pit watchers*) monitor the mud for any changes in weight, viscosity, and temperature. They also look for changes in cutting size, flow rate, and mud level in the tanks. Derrickhands run routine and basic tests to determine the mud's weight, viscosity, and temperature. In some cases, they may also measure the mud's gel strength, wall-building capabilities, and sand content.

Field mud engineers usually do more sophisticated testing. For example, they not only run the tests the derrickhands run, but also they test the mud's pH, liquid and solids content, the presence of contaminants, and its electrolytic properties. A mud that conducts an electric current increases corrosion of the metal components in the hole. Whoever tests the mud also records the measurements in a mud record.

Preparing Mud Samples

To get accurate and useful results from mud tests, mud testers must subject the mud samples to conditions similar to those in the well. To test a newly mixed sample, the mud should be allowed to age for a day or so, so that the dry materials will react and mix properly. Then it should be mixed in a blender or mixer to simulate the effects of circulation in the well.

The tester should age mud at the well's temperature where the well's bottomhole temperature is over the boiling point of water (212°F or 100°C). A mud sample brought in from such a well will have cooled and gelled, so mixing at the proper temperature is necessary to restore the viscosity and other properties the mud exhibits in the well.

Other things to watch include making sure all the air is out of the sample. Also, when taking samples from the active system, be sure to take them from the same place; for example, mud from the shaker is different from mud in the tank.

Density Test

The derrickhand may test the weight and viscosity of drilling mud every 15 minutes to 1 hour, depending on the formation and the complexity of the mud system. Density testing is very important because it is a measure of the mud column's hydrostatic pressure. Density is the weight per unit of volume. Rig personnel may express it as pounds per gallon (ppg), pounds per cubic foot (pcf), or kilograms per cubic centimetre (kg/cm^3).

A *mud balance* measures mud density. A mud balance is a beam balance; it works like the scale in a doctor's office (fig. 31). It has a cup on one end to hold a mud sample and an arm with a sliding weight on it. The arm rests on a fulcrum. The balance weighs a precise volume of mud and automatically divides the weight by the volume to give a direct reading of the mud's weight. To use the balance, fill the cup, move the sliding weight until the arm balances, and read the density from a scale marked on the instrument.

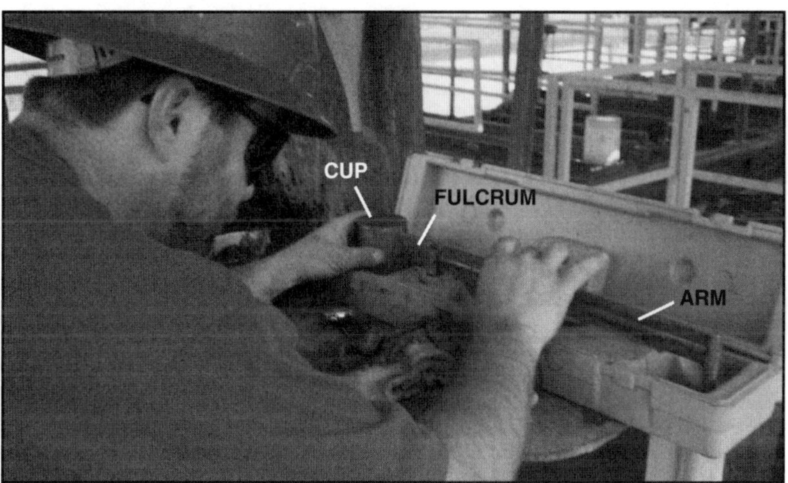

Figure 31. Mud balance (The sliding weight is under the man's right hand.)

Viscosity and Gel-Strength Tests

Two types of instruments measure viscosity: a Marsh funnel and a direct-indicating viscometer. Derrickhands use the Marsh funnel for routine measurements. The direct-indicating viscometer gives a more accurate picture of the mud's flow properties.

Marsh Funnel

A *Marsh funnel* is a funnel about 1 quart (1 litre) in capacity. A mark on the inside top of the funnel indicates the 1 quart (1 litre) level. It comes with a measuring cup (fig. 32). To find a mud's funnel viscosity, fill the funnel with mud to the mark, while holding your finger over the funnel's drain. Remove your finger, start a stop watch, and observe how many seconds it takes for the quart (litre) of mud to flow from the funnel. Finally, record the time on a

Figure 32. Marsh funnel

mud-report form. Express them in terms of seconds per quart (s/qt) or seconds per litre (s/L). Funnel viscosities range from about 30 to 40 seconds for low-solids muds, to about 40 to 50 seconds for high-solids muds, and above 50 seconds for heavier-weight muds. Funnel viscosity of an oil mud is about the same as a water-base mud, but the temperature of the sample is important because the viscosity of oil varies with temperature more than that of water.

To make sure the funnel is working properly, test it periodically with fresh water. A quart (litre) of water at 65°F (18°C) filled to the mark should take 26 seconds to flow through.

Direct-Indicating Viscometer

A *direct-indicating viscometer* measures viscosity and gel strength (fig. 33). It has a weight (bob) inside a rotor (fig. 34). The person running the test fills a cup with mud and lowers the rotor and bob into the cup. Depending on the viscometer, you either turn the rotor with a hand crank or switch on an electric motor. The rotor spins mud that lies inside a space between the rotor and bob.

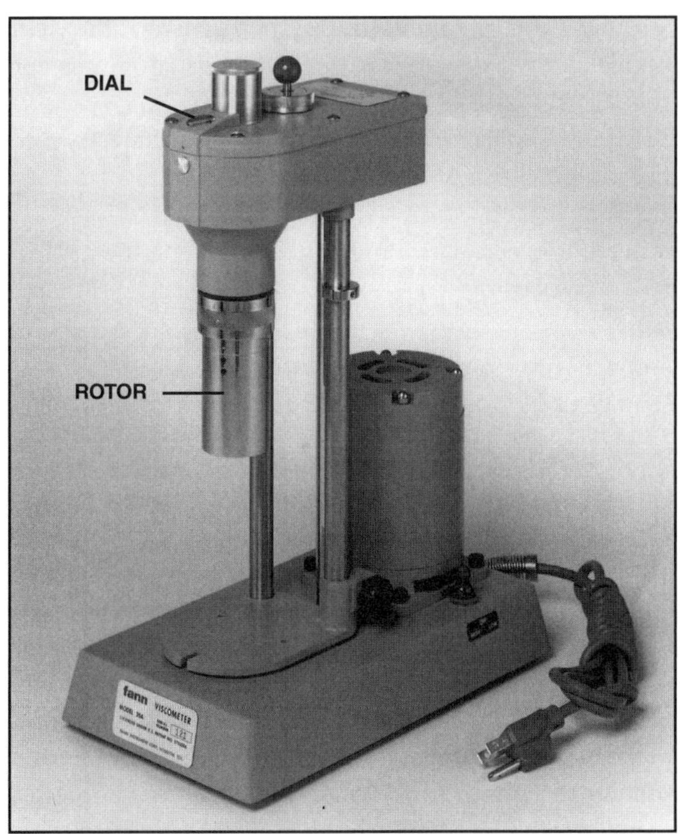

Figure 33. Direct-indicating viscometer

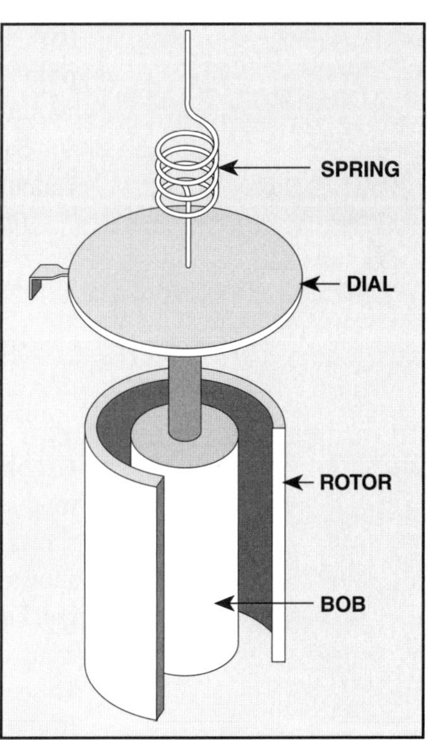

Figure 34. Schematic diagram of the direct indicating viscometer. The deflection in degrees of the bob is read from a graduated scale on the dial.

A spring restrains the movement of the bob, and a dial on top of the viscometer indicates how far the spring moves. The mud's viscosity influences the spring's movement so that thinner muds let the spring move more than thicker muds. The instrument measures a mud's plastic viscosity and the yield point. *Plastic viscosity* is a fluid's resistance to flow because of friction. The *yield point* is a fluid's resistance to flow that results from the attraction between clay particles.

Viscosity. To measure viscosity, fill the cup with mud and place it under the rotor. Adjust the rotor so that the mud in the cup meets a scribe line on the rotor. Set the rotor to turn at 600 revolutions per minute (rpm). (Usually, you set the rpm by pushing down or raising up a button on top of the viscometer. With the button down, the rotor spins at 600 rpm; with the button up, the rotor spins at 300 rpm.) Note the dial and wait for the reading to reach a steady value. Record this value.

Stop the rotor, press the button down, and turn the rotor at 300 rpm. Allow enough time at this speed for the reading to become steady. Record the reading at 300 rpm. The reading at 600 rpm minus the reading at 300 rpm is the plastic viscosity in centipoises (cp). The reading at 300 rpm minus the plastic viscosity is the yield point in pounds per 100 square feet (lb/100ft^2) or kilograms per 10 square metres (kg/10m^2).

Say that the reading is 92 at 600 rpm, and 52 at 300 rpm. Calculate the plastic viscosity and yield point.

Plastic viscosity:

reading at 600 rpm – reading at 300 rpm = plastic viscosity (cp) 92 – 52 = 40 cp

Yield point:

reading at 300 rpm – plastic viscosity (cp) = yield point (lb/100 f^{t2} or kg/10m^2) 52 – 40 = 12 lb/100ft^2 (kg/10m^2)

Gel Strength. To measure gel strength requires two tests on the same mud sample at the same temperature. The first, the 10-second test, shows how quickly the mud gels. The second, the 10-minute test, shows the extent of gelling—how thick it gets after it has a chance to gel. For the 10-second test, lower the viscometer's rotor and bob into a cup of mud and stir it for 10 seconds. Then, let the mud sit still for 10 seconds. Next, stir the mud very slowly, at 3 rpm. Take a reading from the dial on top of the viscometer, which gives the 10-second gel strength in pounds per 100 square feet (kilograms per 10 square metres). For the 10-minute test, stir the mud for 10 seconds as before, but this time let the mud stand for 10 minutes. Then, slowly stir the mud at about 3 rpm to obtain the 10-minute gel strength reading.

Filtration and Wall-Building Tests

Filtration tests measure the relative amount of water in the mud that escapes into a permeable formation and the thickness of the filter cake deposited on the walls of the hole. The test can be made at a low- or high-temperature and pressure.

An instrument called a *filter press* measures filtration rate. A filter press consists of a special cylinder (a cell) that holds a piece of filter paper (special porous paper) and a sample of mud. A pressure attachment at top allows the tester to pressure up the sample to simulate downhole pressure. A graduated cylinder below the cell catches the liquid part of the mud (the filtrate) that pressure forces through the filter paper. The amount of filtrate is in millilitres. The mud solids (wall cake) stay on top of the filter paper. The tester measures the cake's thickness in thirty-secondths of an inch or in millimetres.

The simplest filtration test determines the filtration and wall building rate of mud at low temperature and pressure. To test, put a mud sample and one sheet of filter paper into the cell. Apply pressure (100 psi) from a carbon dioxide or nitrogen cartridge to force the mud through the filter paper. After 30 minutes, release the pressure and read the amount of water that has filtered out of the mud into the graduated cylinder. Also measure the thickness of the solids left on the filter paper in thirty-secondths of an inch (millimetres).

High-temperature tests are run at the actual bottomhole temperature and pressure. A filter press for this purpose (fig. 35) has a pressure source, a cell that can withstand pressures up to 1,000 psi (6,895 kPa), a system for heating the cell, and a frame to hold it.

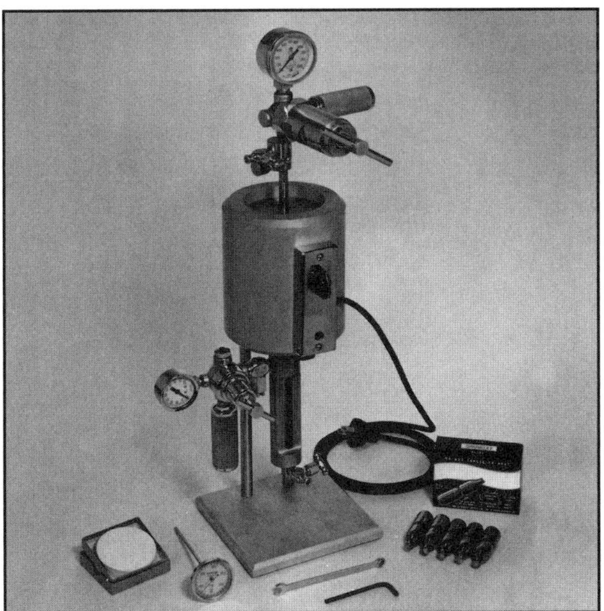

Figure 35. High-pressure/ high-temperature filter press

Measuring Sand Content

A sand content test measures the amount of particles in a mud that cannot pass through a 200-mesh screen. (A 200-mesh screen is a screen that has 200 openings per square inch.) A sand test measures only the size of the particles, not what they are made of. The mud engineer monitors the amount of sand in the mud because sand is a very destructive force on the drill string and bit. A fast-moving circulating mud containing sand abrades metal, like sandblasting, causing it to fail sooner than it should. It may also cause trouble by settling in the hole or increasing the mud weight. Therefore, the operator can add equipment to filter it out if necessary.

The mud tester uses a *screen set* (fig. 36) to determine how much sand is in a mud. It has a 200-mesh sieve inside a collar that fits into a small funnel and a measuring tube to read the amount of sand in percent by volume. To use it, pour a mud sample into the measuring tube to the mark that says, "mud to here." Add water to the "water to here" mark, shake the mixture, and pour it through the screen. Particles smaller than 200 mesh wash through the screen and are disposed of. Then fit the screen with the trapped particles into the funnel and put the end of the funnel into the tube. Wash the sand back into the tube and allow it to settle out. Marks on the tube show how much sand was in the original sample.

Figure 36. Screen set

Solids, Water, and Oil Content

Density, viscosity, gel strength, and filtration rate depend largely on the amount of solids in the mud—the *solids content* or *solids concentration*. The specific gravity of the solids tells the mud engineer the relative amounts of cuttings and weighting material present. This information is especially important for controlling heavy muds. Because viscosity is affected by the solids content, knowing the solids content may explain certain undesirable mud properties and suggest ways to correct them. For example, if the viscosity and the solids content are both high, the operator can probably thin the mud with water rather than with expensive chemicals. And because gel strength and filtration rate are related to the attracting and repelling forces between particles in the mud, the type and amount of these solids affect treatment decisions.

The mud engineer uses distillation to measure the solids content, in an instrument called a *retort*, or a *mud still* (fig. 37). The engineer places a precise volume of mud into a steel container and heats it in the still. The water and oil evaporate and then condense into a graduated cylinder, with marks on the side for measuring the volume. The oil floats on top of the water, so reading the two volumes is easy. Subtracting this measurement from the original volume of mud gives the solids content. Saltwater muds require a correction for the salt content. The mud engineer then compares these measurements with the actual behavior of the mud in drilling.

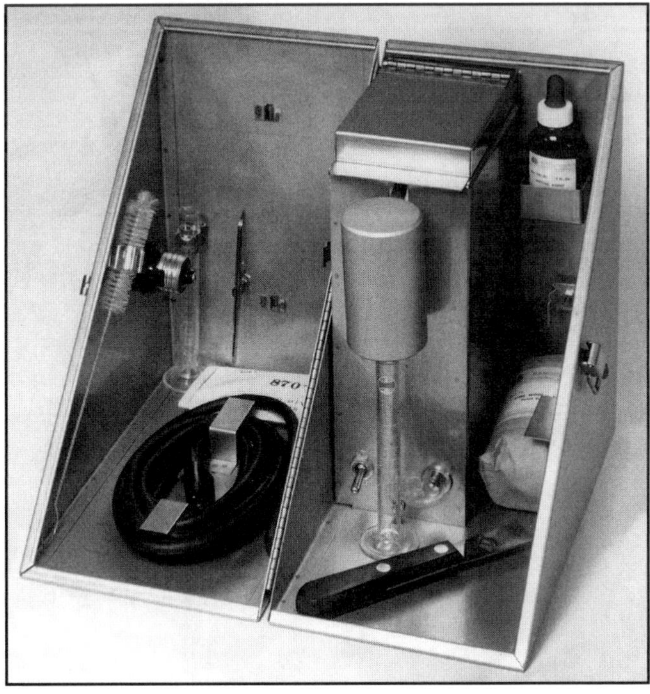

Figure 37. Mud still

Determining pH

Two methods measure pH: the colorimetric method and the electrometric method. The colorimetric method uses a kit that includes pieces of litmus paper and a chart that displays several different colors. Each color corresponds to different pH readings. Litmus paper is chemically treated paper. When placed in the mud, the paper changes color. The color it becomes depends on the mud's pH. The tester matches the color of the paper with the color on the chart and reads the pH. High concentrations of salt may alter the color and make this method unreliable.

The electrometric method uses a pH meter for a very accurate value. The pH meter has two glass electrodes, an amplifier, and a meter. It works on the principle that when a thin membrane of glass separates two solutions with different pH values, an electrical current develops that can be amplified and measured. Usually, an engineer performs an electrometric test.

Other Tests

Mud engineers may do a number of other more sophisticated tests if necessary to determine the presence of contaminants or to help control mud properties. They can do the same tests on water for making up the mud, which may contain dissolved salts that would affect mud treatment.

To summarize—

Main tests done on site

- Weight (density) with a mud balance
- Viscosity and gel strength with a Marsh funnel and direct-indicating viscometer
- Filtration and wall-building capacity with a filter press
- Sand content with a screen set
- Solids, water, and oil content with a mud still
- pH with litmus paper

Treatment of Drilling Mud

The testing that the crew and mud engineer do tells them if and when to treat the mud to ensure that its properties are suitable to the drilling conditions. Changing drilling conditions can make it necessary to change the composition of the mud or change its weight.

Breakover

In many drilling operations, it becomes necessary to change the chemistry of the mud from one type to another. Such a change is referred to as a *conversion*, or *breakover*. Breakover is the point when the properties of the mud actually change. Breakover usually marks a radical change in mud chemistry. During breakover from one type of mud to another, severe "viscosity humps" may occur. For example, crew members may add salt to change a normal bentonite-based mud system to a salt-saturated system. As they add the salt, the viscosity increases to very high levels. As they continue to add salt, however, the mud eventually reaches a breakover point. At the breakover point, the viscosity decreases as crew members continue to add salt. When the viscosity begins to decrease, the mud has broken over from one type of mud system to another.

Reasons for making a breakover include (1) adjusting the mud to maintain a stable wellbore, (2) changing the mud so that it can tolerate more weighting material, (3) altering the mud to drill salt formations, and (4) modifying the mud to reduce its tendency to plug producing zones.

Most operators prefer not to break over a mud in open hole (hole that has no casing in it) because of the high viscosities usually encountered during breakover. With higher viscosities come higher pressures needed to move the mud, which could, in turn, cause a pressure-sensitive formation to fracture. Fractured formations often mean lost circulation and subsequent loss of well control. What is more, high viscosity mud can cause the drill stem to get stuck. Because of such problems, most operators prefer to do a breakover in a cased hole before drilling the next hole section.

Weighting Up

The most important consideration when weighting up is to add the weighting material at a rate that keeps the mud weight constant in the suction tank while circulating. Supervisory personnel need to know the exact weight of the mud going into the hole. Too heavy, and lost circulation can occur; too light, and the well could kick. Careful density measurements of the mud in the suction tank tell whether the weighting material is being added too slowly, too fast, or at the right rate.

A derrickhand in the field can figure how many sacks of barite to add and how fast to add them by approximate calculation. Or a mud engineer can make very accurate calculations. Which method the operator chooses depends on how critical the mix is.

Approximate Calculation (English measurements)

When directed to weight up the mud, the derrickhand must calculate two things: (1) The amount of barite to add and (2) how fast to add it.

Amount of Barite to Add per 100 Barrels. To calculate the amount of barite needed, first calculate how many sacks of barite are needed for each 100 barrels of mud in the circulating system. The way you make the calculation depends on the current mud weight.

1. If the mud weighs less than 12.0 pounds per gallon (ppg), adding 60 sacks of barite will increase each 100 barrels (bbl) of mud by 1 ppg.
2. If the mud weighs more than 12.0 ppg, divide the desired weight by 0.2 to find the number of sacks of barite needed to increase each 100 barrels of mud by 1 ppg.

$$\frac{\text{desired weight, ppg}}{0.2} = \text{number of sacks per 100 bbl of mud}$$

Volume of Mud to Be Treated. The next step in determining how much barite to add is to calculate the volume of mud in the system.

3. Total volume of mud in the system includes the volume of mud in the hole plus the volume of mud in the mud tanks.

   total volume (bbl) = hole volume (bbl) + tank volume (bbl)

   a. Calculate the volume of mud in the hole approximately or accurately. For an approximate calculation, first determine the approximate volume of the hole per 1,000 feet of depth. Simply squaring the diameter of the hole in inches gives the approximate hole volume per 1,000 feet of depth.

$$\text{hole volume (bbl) per 1,000 feet of depth} = \text{diameter of hole (inches)}^2$$

Next, multiply the hole volume per 1,000 feet by the depth in thousands of feet to find the total volume of the hole.

$$\text{total hole volume (bbl)} = \text{hole volume per 1,000 feet (bbl)} \times \text{depth of hole (1,000 feet)}$$

b. For a more accurate calculation, first square the diameter of the hole.

$$\text{hole volume (bbl) per 1,000 feet of depth} = \text{diameter of hole (inches)}^2$$

Then divide the result by 1,029 and multiply that number by the depth of the hole in feet.

$$\text{total hole volume (bbl)} = \frac{\text{hole volume per 1,000 feet (bbl)} \times \text{depth of hole (feet)}}{1{,}029}$$

c. Calculate the volume of each tank by multiplying each tank's length by its width by the depth of the mud. (The depth of the mud remains constant all the time.) The answer is in cubic feet (ft^3).

$$\text{volume of tank (ft}^3\text{)} = \text{length (ft)} \times \text{width (ft)} \times \text{mud depth (ft)}$$

d. Add the volumes of all the active tanks together.

$$\text{total tank volume (ft}^3\text{)} = \text{volume of tank one (ft}^3\text{)} + \text{volume of tank two (ft}^3\text{), etc.}$$

e. Convert the volume in cubic feet to barrels so that the units of volume in the tanks match the units of volume in the hole. Each cubic foot of mud equals about 5.6 barrels, so divide by 5.6.

$$\text{total hole volume (bbl)} = \frac{\text{volume of tanks (ft}^3\text{)}}{5.6}$$

f. Finally, add total hole volume to the total tank volume to find the volume of mud in the circulating system.

$$\text{total volume of mud (bbl)} = \text{hole volume (bbl)} + \text{tank volume (bbl)}$$

4. Now multiply the results of step 1 or 2 by the results of step 3f and divide by 100 to find how much barite to add.

$$\text{total barite (sacks)} = \frac{\text{barite per 100 bbl mud (sacks)} \times \text{volume of mud (bbl)}}{100}$$

Rate of Addition. To figure how fast to add the barite, you need to calculate the *cycle time*, the time needed for the mud to make one complete circulating cycle from the suction intake at the pump to the bottom of the hole and back to the pump's suction. Cycle time is important because it is desirable to add all the barite during one circulating cycle.

5. First calculate the pump output in barrels per minute (bbl/min) using one of these equations:

$$\text{pump output (bbl/min)} = \text{pump output (gal/min)} \times 0.024$$

or

$$\text{pump output (bbl/min)} = \text{bbl/stroke} \times \text{strokes/min}$$

6. Calculate the cycle time by dividing the results of step 3f by the results of step 5.

$$\text{total barite (sacks)} = \frac{\text{total volume of mud (bbl)}}{\text{pump output (bbl/min)}}$$

7. Figure out how many sacks to add per minute by dividing the number of sacks needed (step 4) by the cycle time (step 6).

$$\text{total barite (sacks)} = \frac{\text{total volume of sacks}}{\text{cycle time (min)}}$$

Most hoppers can mix 5 to 10 sacks per minute without too much difficulty. To add barite as fast as 15 sacks per minute, the rig requires more than one hopper, all in good condition, and bulk barite storage tanks. If bulk barite is not available, crew members must proceed at a slower rate. To mix at a slower rate, the driller can circulate at a slower rate. The slower pumping rate allows crew members to raise the mud weight over two or more circulation cycles.

Sample Problem. Let's solve a problem using the steps above. You must raise the mud weight from 18.0 to 19.0 ppg, and you know the following about the system:

- The hole has a diameter of 9⅞ (or 9.875) inches and is 10,000 feet deep.
- The system has two mud tanks. Each measures 8 feet by 20 feet and the mud is 7 feet deep.
- The pump output is 417 gallons per minute.
- The weighting material is barite.

1. Because the mud weight is more than 12.0 ppg, begin with step 2.
2. To calculate the number of sacks of barite per 100 barrels of mud

$$\frac{19.0 \text{ ppg}}{0.2} = 95 \text{ sacks per 100 bbl mud}$$

3a. For an approximate calculation of hole volume, round off the hole's diameter of 9⅞ inches to 10 inches.

$$\text{hole volume per 1,000 feet of depth} = 10^2 = 100 \text{ bbl}$$

The hole is 10,000 feet deep, so multiply this number by 10.

$$\text{total hole volume} = 100 \text{ bbl} \times 10{,}000 \text{ feet} = 1{,}000 \text{ bbl}$$

3b. For an accurate calculation, do not round off the hole diameter.

$$\text{hole volume per 1,000 feet depth} = 9.875^2 = 97.516$$

Multiply by the depth in feet and divide by 1,029

$$\text{total hole volume} = \frac{97.516 \times 10{,}000}{1{,}029} = 948 \text{ bbl}$$

3c. Calculate the volume of each tank.

$$\text{volume of each tank} = 8 \text{ ft} \times 20 \text{ ft} \times 7 \text{ ft} = 1{,}120 \text{ ft}^3$$

3d. Add the volumes of the two tanks.

$$\text{total tank volume} = 1{,}120 \text{ ft}^3 + 1{,}120 \text{ ft}^3 = 2{,}240 \text{ ft}^3$$

3e. Convert cubic feet to barrels.

$$\text{total tank volume} = \frac{2{,}240 \text{ ft}^3}{5.6} = 400 \text{ bbl}$$

3f. Calculate the total volume of mud in the circulating system by adding the results of steps 3a (or 3b) and 3e.

$$\text{total volume of mud} = 400 + 1{,}000 = 1{,}400 \text{ bbl}$$

4. Calculate the total number of sacks of barite to add

$$\text{total hole volume} = \frac{95 \times 1{,}400}{100} = 948 \text{ bbl}$$

Now you know to add 1,330 sacks of barite to raise the weight of the mud to 19.0 ppg. Next figure how fast to add it.

5. Convert the pump output in gallons per minute to barrels per minute using the equation in step 5.

$$\text{pump output} = 417 \text{ gal/min} \times 0.024 = 10 \text{ bbl/min}$$

6. Now calculate the cycle time using the results of steps 3f and 5.

$$\text{cycle time (min)} = \frac{1{,}400 \text{ ft}^3}{10 \text{ bbl/min}} = 140 \text{ minutes}$$

You will need to add all the barite over a period of 140 minutes.

7. Figure out how many sacks to add per minute by dividing the number of sacks needed (step 4) by the cycle time (step 6).

$$\text{sacks per minute} = \frac{1{,}330 \text{ sacks}}{140 \text{ min}} = 9.5 \text{ sacks per minute}$$

Approximate Calculation (SI Metric Measurements)

When directed to weight up the mud, the derrickhand must calculate two things: (1) The amount of barite to add and (2) how fast to add it.

Amount of Barite to Add. To calculate the amount of barite needed, first figure how many kilograms (kg) of barite are needed for each cubic metre (m^3) of mud in the circulating system.

1. Use the following equation when adding barite:

$$\text{amount of barite per m}^3 \text{ of mud} = \frac{4{,}250 \text{ (desired density, kg/m}^3 - \text{initial density, kg/m}^3}{4{,}250 - \text{desired density, kg/m}^3}$$

Volume of Mud to Be Treated. Next determine how many cubic metres of mud are in the system.

2. Total volume of mud in the system includes the volume of mud in the hole plus the volume of mud in the mud tanks.

$$\text{total volume} = \text{hole volume} + \text{tank volume}$$

   a. Calculate the approximate volume of the hole by squaring the diameter of the hole and multiplying by the depth. Because the hole diameter is in millimetres (mm), convert first to metres (m) by dividing by 1,000.

$$\text{hole diameter (m)} = \frac{\text{hole diameter (m)}}{1{,}000}$$

   b. Calculate the volume of each tank by multiplying each tank's length by its width by the depth of the mud. (The depth of the mud remains constant all the time.)

$$\text{volume of tank (m}^3) = \text{length (m)} \times \text{width (m)} \times \text{depth (m)}$$

c. Add the volumes of all the active tanks together.

$$\text{total tank volume (m}^3) = \text{volume of tank 1 (m}^3) + \text{volume of tank 2 (m}^3)\text{, etc.}$$

d. Finally, add total hole volume to total tank volume to find the volume of mud in the circulating system.

$$\text{total volume of mud (m}^3) = \text{hole volume (m}^3) + \text{tank volume (m}^3)$$

3. Now multiply the results of step 1 by the results of step 2d to figure how much barite to add.

$$\text{total barite} = \text{barite per m}^3 \text{ mud (kg)} \times \text{volume of mud (m}^3)$$

4. Convert kilograms to sacks. If a sack contains 40 kilograms barite, divide the result of step 3 by 40.

$$\text{sacks of barite} = \frac{\text{kilograms of barite}}{\text{40 kg/sack}}$$

Rate of Addition. To figure how fast to add the barite, calculate the *cycle time*, the time needed for the mud to make one complete circulating cycle from the suction intake at the pump to the bottom of the hole and back to the pump's suction. Cycle time is important because it is desirable to add all the barite during one circulating cycle.

5. First calculate the pump output in cubic metres per minute (m^3/min) using one of these equations:

$$\text{pump output (m}^3\text{/min)} = \text{pump output (litres/min)} \times 0.001$$

or

$$\text{pump output (m}^3\text{/min)} = \text{litres/stroke} \times \text{strokes/min} \times 0.001$$

6. Calculate the cycle time by dividing the results of step 2d by the results of step 5.

$$\text{cycle time (min)} = \frac{\text{total volume of mud (m}^3)}{\text{pump output (m}^3\text{/min)}}$$

7. Figure out how many sacks to add per minute by dividing the number of sacks needed (step 4) by the cycle time (step 6).

$$\text{sacks per minute} = \frac{\text{total number of sacks}}{\text{cycle time (min)}}$$

Sample Problem. Let's solve a problem using the steps above. You must raise the mud weight from 1,080 to 1,320 kilograms per cubic metre, and you know the following about the system:

- The hole has a diameter of 250 millimetres and is 3,000 metres deep.
- The system has two mud tanks. Each measures 2.5 metres by 6 metres and the mud is 2 metres deep.
- The pump output is 1,580 litres per minute.
- The weighting material is barite.

1. Calculate the amount of barite needed for each cubic metre of mud in the system.

$$\text{amount of barite per m}^3 \text{ of mud} = \frac{4{,}250\,(1{,}320 \text{ kg/m}^3 - 1{,}080 \text{ kg/m}^3)}{4{,}250 - 1{,}320 \text{ kg/m}^3} = 348 \text{ kg/m}^3$$

2a. Calculate the approximate volume of the hole after converting its diameter from millimetres to metres.

$$\text{hole diameter} = \frac{250 \text{ mm}}{1{,}000} = 0.25 \text{ m}$$

$$\text{total hole volume} = (0.25\text{m})^2 \times 3{,}000 \text{ m} = 188 \text{ m}^3$$

2b. Calculate the volume of each tank.

$$\text{volume of each tank} = 2.5 \text{ m} \times 6 \text{ m} \times 2 \text{ m} = 30 \text{ m}^3$$

2c. Add the volumes of the two tanks.

$$\text{total tank volume} = 30 \text{ m}^3 + 30 \text{ m}^3 = 60 \text{ m}^3$$

2d. Finally, add total hole volume to the total tank volume to find the volume of mud in the circulating system.

$$\text{total volume of mud} = 188 \text{ m}^3 + 60 \text{ m}^3 = 248 \text{ m}^3$$

3. Now multiply the results of step 1 by the results of step 2d to figure how much barite to add.

$$\text{total barite} = 348 \text{ kg/m}^3 \times 248 \text{ m}^3 = 86{,}304 \text{ kg}$$

4. Convert kilograms to sacks.

$$\text{sacks of barite} = \frac{86{,}304 \text{ kg}}{40 \text{ kg/sack}} = 2{,}158 \text{ sacks}$$

Now you know to add 2,158 sacks of barite to raise the weight of the mud to 1,320 kilograms per cubic metre. Next figure how fast to add it.

5. First calculate the pump output in cubic metres per minute using the equation in step 5:

$$\text{pump output (m}^3\text{/min)} = 1{,}580 \text{ litres/min} \times 0.001 = 1.58 \text{ m}^3$$

6. Now calculate the cycle time by dividing the results of step 2d by the results of step 5.

$$\text{cycle time (min)} = \frac{248 \text{ m}^3}{1.58 \text{ m}^3} = 157 \text{ minutes}$$

You will need to add all the barite over a period of 157 minutes.

7. Figure out how many sacks to add per minute by dividing the number of sacks needed (step 4) by the cycle time (step 6).

$$\text{sacks per minute} = \frac{2{,}158 \text{ sacks}}{157 \text{ min}} = 14 \text{ sacks per minute}$$

Calculating Increased Volume

Adding weighting material also increases the volume of mud, which the derrickhand needs to take into consideration. The total mud volume must not exceed the holding capacity of the system.

Calculation Using English Units

The addition of each 100 sacks of barite increases mud volume by 6.7 barrels. So to calculate how much the total volume will increase, divide the number of sacks to be added by 6.7.

$$\text{volume increase (bbl)} = \frac{\text{number of sacks of barite added}}{6.7}$$

Using the sample problem given earlier, how much will the total mud volume increase?

$$\text{volume increase} = \frac{1{,}330 \text{ sacks}}{6.7} = 198.5 \text{ bbl}$$

Calculation Using SI Metric Units

To calculate the volume increase, divide the amount of barite added by the constant of 4,250.

$$\text{volume increase (m}^3\text{)} = \frac{\text{kg barite added}}{4{,}250}$$

Using the sample problem given earlier, how much will the total mud volume increase?

$$\text{volume increase (m}^3\text{)} = \frac{86{,}304 \text{ kg barite}}{4{,}250} = 20.3 \text{ m}^3$$

Using Tables Table 4 shows the number of 100-pound sacks of barite required to raise the weight of 100 barrels of mud (of a given initial weight in ppg) to a desired weight (in ppg). It also shows the number of barrels of water to add to 100 barrels of mud to reduce its weight. Table 5 shows an equivalent metric table to raise the mud weight. Tables make it easy to estimate how much barite to add to weight up a mud a certain amount.

Table 4
Mud-Weight Adjustment with Barite or Water

Initial Mud Weight (ppg)	Desired Mud Weight (ppg)																	
	9.5	10.0	10.5	11.0	11.5	12.0	12.5	13.0	13.5	14.0	14.5	15.0	15.5	16.0	16.5	17.0	17.5	18.0
9	29	59	90	123	156	192	229	268	308	350	395	442	490	542	596	653	714	778
9.5		29	60	92	125	160	196	234	273	315	359	405	452	503	557	612	672	735
10	43		30	61	93	128	164	201	239	280	323	368	414	464	516	571	630	691
10.5	85	30		31	62	96	131	167	205	245	287	331	376	426	479	531	588	648
11	128	60	23		31	64	98	134	171	210	251	294	339	387	437	490	546	605
11.5	171	90	46	19		32	66	101	137	175	215	258	301	348	397	449	504	562
12	214	120	69	37	16		33	67	103	140	179	221	263	310	357	408	462	518
12.5	256	150	92	56	32	14		34	68	105	144	184	226	271	318	367	420	475
13	299	180	115	75	48	27	12		34	70	108	147	188	232	278	327	378	432
13.5	342	210	138	94	63	41	24	11		35	72	111	150	194	238	286	336	389
14	385	240	161	112	76	54	36	21	10		36	74	113	155	199	245	294	345
14.5	427	270	185	131	95	68	48	32	19	9		37	75	116	159	204	252	303
15	470	300	208	150	110	82	60	43	29	18	8		37	77	119	163	210	259
15.5	513	330	231	169	126	95	72	54	39	26	16	8		39	79	122	168	216
16	556	360	254	187	142	109	84	64	48	35	24	15	7		40	81	126	172
16.5	598	390	277	206	158	123	96	75	58	44	32	23	14	7		41	84	129
17	641	420	300	225	174	136	108	86	68	53	40	30	21	13	6		42	86
17.5	684	450	323	244	189	150	120	96	77	62	49	38	28	20	12	6		43
18	726	480	346	262	205	163	132	107	87	71	57	45	35	26	18	12	5	

The lower left half of this table shows the number of barrels of water that must be added to 100 bbl of mud to produce desired *weight reductions*. To use this portion of the table, locate the initial mud weight in the vertical column at the left, then locate the desired mud weight in the upper horizontal row. The number of barrels of water to be added per 100 bbl of mud is read directly across from the initial weight and directly below the desired mud weight. For example, to reduce an 11 ppg mud to a 9.5 ppg mud, 128 bbl of water must be added for every 100 bbl of mud in the system.

The upper right half of this table shows the number of sacks of barite that must be added to 100 bbl of mud to produce desired *weight increases*. To use this portion of the table, locate the initial mud weight in the vertical column to the left, then locate the desired mud weight in the upper horizontal row. The number of sacks of barite to be added per 100 bbl of mud is read directly across from the initial weight and directly below the desired mud weight. For example, to raise an 11 ppg mud to 14.5 ppg, 251 sacks of barite must be added per 100 bbl of mud in the system.

For example, table 4 shows that raising 13.0 ppg mud to 15.0 ppg requires 147 sacks of barite per 100 barrels of mud. Similarly, table 5 shows that raising density from 1,300 to 1,500 kg/m^3 requires about 300 kg of barite per m^3 of mud.

Table 5
Mud Weight-up Chart
SI Metric

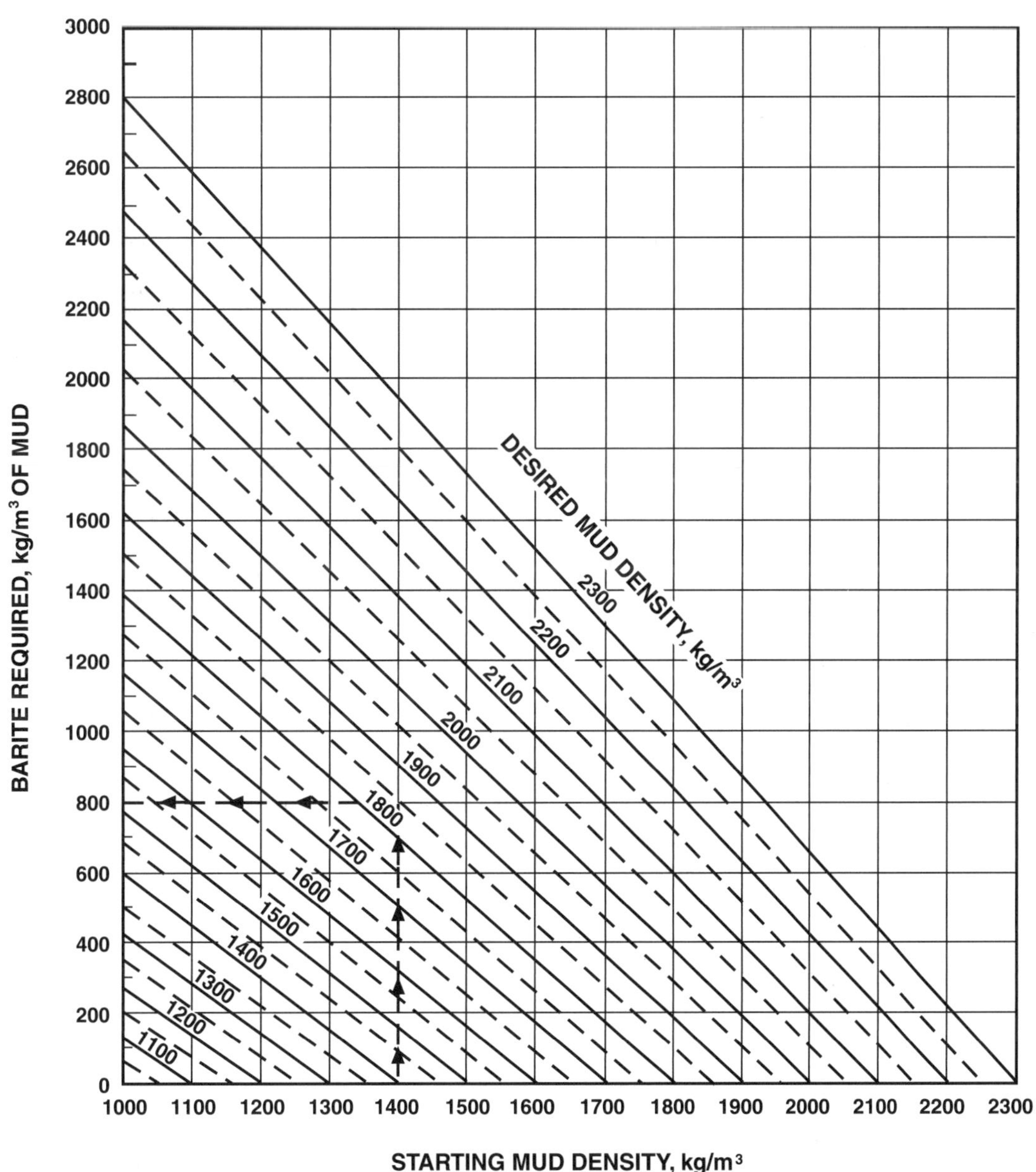

Calculating Hydrostatic Pressure

A related calculation that the derrickhand makes is to determine hydrostatic pressure downhole. To make the calculation, measure the weight of the mud with a mud balance, then multiply the weight by the depth of the fluid column (depth of the hole). Depending on whether the balance measures mud weight in pounds per gallon (ppg), pounds per cubic foot (pcf), or kilograms per metre (kg/m), choose one of the following equations to calculate the hydrostatic pressure:

$$\text{hydrostatic pressure (psi)} = \text{depth (ft)} \times \text{mud weight (ppg)} \times 0.052$$

$$\text{hydrostatic pressure (psi)} = \text{depth (ft)} \times \text{mud weight (pcf)} \times 0.00695$$

$$\text{hydrostatic pressure (psi)} = \frac{\text{depth (m)} \times \text{mud weight (kg/m}^3\text{)}}{100}$$

Example Calculations

Determine the hydrostatic pressure in psi at the bottom of a 9,623-foot hole that is full of mud that weighs 12.2 ppg. Using the equation, hydrostatic pressure (psi) = depth (ft) × mud weight (ppg) × 0.052, the solution is—

$$\text{hydrostatic pressure} = 9{,}623 \times 12.2 \times 0.052 = 6{,}105 \text{ psi.}$$

Determine the hydrostatic pressure in psi at the bottom of a 9,623-foot hole that is full of mud that weighs 91.26 pcf. Using the equation, hydrostatic pressure (psi) = depth (ft) × mud weight (pcf) × 0.00695, the solution is—

$$\text{hydrostatic pressure} = 9{,}623 \times 91.26 \times 0.00695 = 6{,}104 \text{ psi.}$$

Determine the hydrostatic pressure in kPa at the bottom of a 2,933-metre hole that is full of mud that weighs 1,462 kg/m^3. Using the equation, hydrostatic pressure (kPa) = depth (m) × mud weight (kg/m^3) ÷ 100, the solution is—

$$\text{hydrostatic pressure} = 2{,}933 \times 1{,}462 = 4{,}288{,}046 \div 100 = 42{,}880 \text{ kPa.}$$

Water-Back

Occasionally, the operator wants to reduce the mud weight rather than increase it. For example, a heavier mud may be needed to drill a high-pressure formation. Then, after casing is set, the casing seals off the high-pressure formation. If the formations below the casing have normal pressure, the operator may then wish to decrease the mud weight.

In water-base muds, adding water is the usual way to reduce mud weight. Adding water to reduce mud weight is one meaning of the term *water-back*. (Water-back also refers to adding water in order to decrease the solids content of the mud.)

Calculating Water-Back (Using English Units)

The following formula approximates the volume of water needed to reduce mud weight:

$$x = \frac{V(W_1 - W_2)}{W_2 - 8.34}$$

where

x = bbl of water needed
V = original volume of mud, bbl
W_1 = initial mud weight, ppg
W_2 = desired mud weight, ppg.

For example, if the total volume of mud in the system is 1,000 bbl and you want to reduce the mud weight from 12.0 ppg to 11.0 ppg, then calculate as follows:

$$x = \frac{1{,}000\ (12.0 - 11.0)}{11.0 - 8.34} = 376 \text{ bbl of water}$$

Calculating Water-Back (Using SI Units)

The following formula approximates the volume of water needed to reduce mud weight:

$$x = \frac{V(D_1 - D_2)}{D_2 - 1{,}000}$$

where

x = m^3 of water needed
V = original volume of mud, m^3
D_1 = initial mud weight, kg/m^3
D_2 = desired mud weight, kg/m^3.

For example, if the total volume of mud in the system is 143 m^3 and you want to reduce the mud weight from 1,320 kg/m^3 to 1,080 kg/m^3, then calculate as follows:

$$x = \frac{143\,(1{,}320 - 1{,}080)}{1{,}080 - 1{,}000} = 429 \text{ m}^3 \text{ of water}$$

These formulas do not consider the effect of solids settling because of decreased viscosity. When the crew adds large amounts of water, they must also add materials to increase the viscosity to prevent the weighting material from settling out of the mud. They add them slowly through the hopper to avoid plugging the hopper.

Factors Affecting Mud Performance and Cost

The drilling crew should be aware of factors that affect the performance and cost of drilling mud. An economical drilling operation keeps the contractor and rig crews working. Rig owners and crew members should keep in mind the following points about the circulating system and equipment.

- The mud tanks should provide the necessary surface volume and be set up to allow the mud to move freely from one tank to another.
- The mud mixing equipment, such as centrifugal pumps, hoppers, and agitators should be large enough for the system. Also, the crew must maintain them periodically to ensure that they are in top working order.
- The rig owner should install shale shakers, degassers, desanders, desilters, mud cleaners, and centrifuges to reduce mud cost.
- The rig owner should provide good mud storage facilities, which protect the mud from weather and eliminates spoilage.
- Crew members should add water to water-base muds continuously while the pumps circulate the mud. Adding water replaces that lost by surface evaporation and downhole filtration. Crew members should add it to a tank and not on the shale shaker. Water on the shaker could wash cuttings through the shaker's screen.
- Mud engineers should remain alert to changes in the mud that may indicate that the mud needs treatment to offset downhole problems. Further, they should note any changes in surface volume.
- Operators should pay close attention to the type of mud being used for a particular section of hole. They should use the best type of mud for the formation, based on efficiency and cost.

To summarize—

Breakover or conversion

- Reasons for making a breakover include
  - Maintaining a stable wellbore
  - Providing a mud that will tolerate higher weight
  - Drilling salt formations
  - Reducing the plugging of producing zones

Weight up

- The derrickhand calculates approximately how much weighting material to add and how fast to add it, ideally to complete the addition in one circulation cycle.
- Adding weighting material increases the volume of mud in the system.

Water-back

- The derrickhand calculates how much water to add in order to reduce viscosity.

Safe Handling of Muds and Additives

Some of the materials that go into a drilling mud are dangerous if inhaled or if they contact the skin and clothing. Oil-based materials are hazardous mainly because of the possibility of fire and explosion. The crew should always be aware of the dangers, take steps to prevent accidents and injuries, and know the best first-aid treatment in case of an accident.

Storage and Handling of Mud Materials

Most mud materials arrive at the rig site in dry form, generally in sacks. Exceptions are some liquid chemicals, the liquid components of oil-base mud, and new or salvaged liquid mud. Bentonite, clays, starches, CMC, lignins, lignosulfonates, polymers, and other substances usually come in multilayer paper bags weighing 50 to 100 pounds (23 to 45 kilograms). Barite may be in 100-pound (45-kilogram) bags, but often it is supplied in bulk form from trucks or boats. Rig owners can also obtain bentonite in bulk form.

Sacked Material

Usually trucks transport sacked materials to the drilling rig location. There the crew stores the sacks in a structure called a *mud house* or mud room for protection from the weather (fig. 38). The crew usually leaves the sacks on the wooden pallets they are delivered on so that any water on the floor of the mud house does not directly contact the sacks. As an additional precaution, they may cover the sacks with a heavy-duty plastic tarp. The crew should stack sacks in straight, even rows, keeping the various materials separate for easy handling and counting. Use broken sacks first.

Figure 38. Sacks of material in a mud house (Courtesy of Nabors Drilling USA)

The rig-up crew may mount the mud house at ground level or on an elevated foundation. In either case, it is important for crew members to protect the sacked materials so that water on the ground does not damage the sacks on the bottom. Rig-up crews should locate the mud house where delivery trucks can easily get to it. Also, they should place it close to the mixing hopper and on the same level to make moving the sacks as easy as possible.

Bulk Material

Handling barite in bulk form has advantages over the sacked form. Less labor is needed each time the material is moved because machinery does the work. Adding the barite to the mud stream is faster. Broken sacks are not a problem. And handling, storing, and mixing operations are cleaner.

Transferring and Storing Bulk Materials. On land rigs, the crew transfers bulk barite from the transport truck into a manhole on top of a storage bin, called a *P-tank* (fig. 39). They make the transfer through a hose using pressurized air. An air compressor sends pressurized air into the bottom of the truck's tank through a porous membrane, the *air slide*. The air works its way through the barite powder to the top of the truck's tank, which causes the powder to behave like a fluid. Air pressure pushes the fluidized barite to the discharge pipe and out of the hose into the P-tank. An air slide can move barite at 30 to 50 tons (31 to 51 tonnes) per hour for several hundred feet (up to a hundred metres).

Barges and supply boats use similar tanks to transport bulk materials, including cement, along canals or to offshore locations. On offshore rigs, a special pressure tank pushes aerated barite from the barge or boat to the P-tanks on the elevated platforms, which may be 100 feet (30 metres) above the water. To move the material for mixing, the derrickhand opens certain valves and the rig's compressed air system drives the powder into the mixing hopper.

On land rigs, P-tanks have capacities of 500 to 1,000 sacks. The bottom of the tank is tapered and has an air slide and gate discharge opening. The fluidized barite flows directly by gravity into a mixing hopper beneath the tank. Bulk tanks can mix the barite into the mud very quickly, sometimes faster than the mud pumps can handle it. The actual rate of mixing depends on the power of the mixing pump and how fast the material flows through the hopper.

Rig personnel can determine the amount of barite that the P-tank system has added in one of two ways: they can measure the increase in weight of the mixed mud and the volume treated, or they can check the level of the barite remaining in the P-tank.

When a drilling job is finished, the crew transfers leftover bulk material in the P-tanks back to a transport truck. They use a blower and a special aeration unit called an *air bazooka* to aerate the powder in the bin and move it through a hose.

Figure 39. Two P-tanks on a land rig

Handling Chemicals

When the operator brings new chemicals to the site, crew members receive instructions on how to handle and store these chemicals. Manufacturers provide *material safety data sheets (MSDS)* that outline what precautions to take. Where required, crew members should use personal protective equipment (PPE), such as rubber gloves, aprons, face shields, and respirators.

Mixing the Chemicals

Rig owners usually provide a small tank or *chemical barrel* for crew members to mix chemicals into the mud system. Usually, they mount it next to a mud tank (fig. 40). An adequate installation provides easy access and sturdy mounting. The rig owner should make sure that the top of the barrel is no more than waist high. Persons using the barrel should always be able to keep their face and upper body in the clear. Otherwise, chemicals could splash out of the barrel and onto the torso or face. Even though persons should use personal protective equipment to protect against such splashes, it is better to avoid them as much as possible.

Figure 40. Equipment for chemical treatment of mud

Mixing dry chemicals through the mud hopper is generally not hazardous. Note, however, that crew members should never use the hopper to mix caustic soda. It is too likely for the hopper to eject the soda out of the top. As a result, caustic soda could easily splash onto a crew member. Crew members should, therefore, carefully add caustic soda through the chemical barrel. Also, they should never add water to dry caustic soda. It boils up and splashes onto anyone nearby.

Also, if several crew members are adding chemicals through the hopper at the same time, use teamwork and be concerned about protecting each other. Be careful not to clog the hopper by pouring too fast, as a clogged mud hopper can cause materials to splash up. And, of course, all personnel should be wearing appropriate personal protective equipment.

Caustic Soda

As mentioned earlier, *caustic soda*, or *sodium hydroxide*, is a dangerous chemical to handle. In fact, it is one of the most dangerous materials at the drilling site. It comes in the form of beads or pellets in a paper bag. Caustic soda, also shortened to *caustic*, is a strong alkali that burns the skin, especially wet skin. For this reason, the eyes, which are moist, are particularly vulnerable to it. Muds that are heavily treated with caustic also burn the skin, either on direct contact or after soaking through clothing. Caustic on the skin can result in the need for a skin graft, so take no chances with it.

Also mentioned earlier was the warning to never use the hopper to add caustic soda directly to mud. Instead, mix the caustic with water or oil first in a special chemical tank. Careful handling is important even when using a chemical tank because caustic in water generates heat that can cause the tank to boil over. Never put caustic in the tank first and then add water. This method can cause the powder to splash up onto the body or face. Always add the caustic to water already in the tank. Always wear goggles, rubber gloves, face shields, body and torso protection, and respirators when handling caustic.

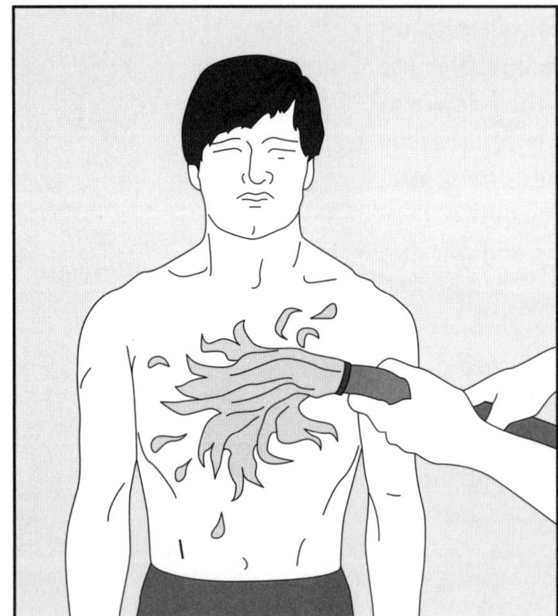

Figure 41. First aid for chemical burns of the skin

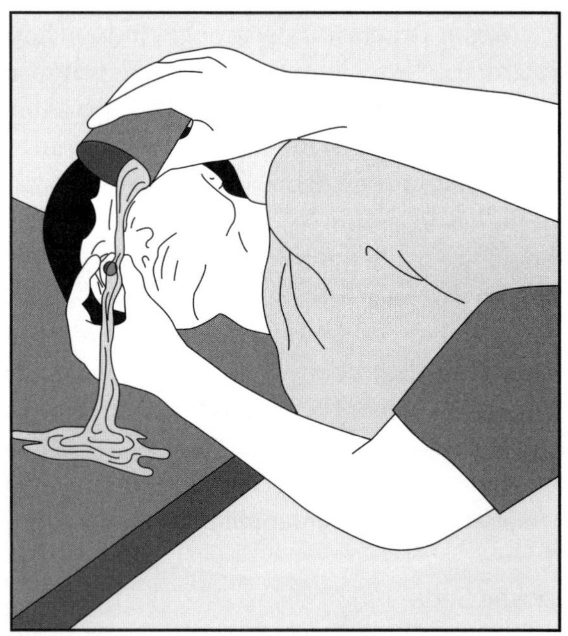

Figure 42. First aid for chemical burns of the eye

If you are exposed to caustic and receive a burn, the first rule is to wash the burn with a flood of water (fig. 41). Afterwards, a wash with vinegar is helpful. Vinegar is mildly acidic and neutralizes any remaining alkalinity on the affected area. Be careful of muddy clothing. Caustic can slowly burn through clothing to the skin, so wash treated mud off your clothing. Eye burns need prompt washing with a lot of water (fig. 42), followed by application of boric acid ointment or castor oil. Then get to a doctor as quickly as possible.

Starch Preservatives

Starch preservatives in treated mud have a formaldehyde base, and often other bactericides as well. These are poisonous if swallowed, and their fumes and dust can seriously affect the lungs and eyes. Use gloves, face shields, body protection, and respirators when handling them.

Handling Oil Muds

Using oil-base or invert-oil muds requires special safety precautions not necessary with water-base muds. First, oil muds present a risk of fire because the oil is flammable. The type of oil in these muds has a high flash point (it ignites at a high temperature) to reduce fire hazard, and the muds contain from 5 to 50 percent water, which further reduces the fire hazard. Most oil muds contain enough solids and water to keep them from burning. Still, do not assume that oil muds are fireproof. Always treat oil muds as if they could easily ignite. Electrical equipment on the rig should be explosion-proof to keep sparks away from the mud. Smoke only in designated areas on the rig site. The risk of fire increases if the mud overheats, becomes contaminated with crude oil (which has a lower flash point), or formation gas enters the mud. In such cases, take extra precautions to avoid fires.

Second, oil muds are slippery. Keep mud cleaned from rig components and equipment to prevent falls and slipping. Oil muds are also hard on equipment. The oil damages rubber parts such as hoses and gaskets, so the operator will use parts made of synthetic, oil-resistant rubber when drilling with oil muds. Oil muds are hazardous to the environment as well, and closely regulated. Do not allow them to flow anywhere except where they belong—the tanks, downhole, or to special disposal areas.

Finally, because these muds are expensive, keep drilling fluid losses to a minimum, and avoid contaminating them with water.

To summarize—

- Store sacked mud additives in the mud house, and bulk materials in P-tanks, to protect them from weather.
- Wear protective gear (personal protective equipment, or PPE) and follow the precautions listed on material safety data sheets (MSDS) when handling caustic, starch preservatives, and other chemicals.
- Know appropriate first-aid treatment in case of accidents.
- When drilling with oil muds, use explosion-proof rig equipment, and smoke only in designated areas. Keep mud off equipment and the rig floor.

Pumps on the Rig

The mud pumps, or slush pumps, are the most important pieces of equipment in a circulation system that uses liquid drilling fluid. If they break down during drilling, the operation comes to a halt. They must therefore be reliable. One pump might be sufficient under some conditions if the other stopped working, but safety requires at least two working pumps at all times. Therefore, mud pumps are extremely sturdy, capable of handling heavy loads, and can tolerate abrasive fluids. Most drilling rigs also have several small auxiliary pumps, usually centrifugal pumps, to move mud in and around the circulating system. Generally, these small pumps do not develop pressures higher than about 150 psi (1,034 kPa).

The derrickhand, working with other crew members, is responsible for making sure that the pumps work efficiently, maintaining them, and repairing them when needed.

Rotary rigs use two types of pumps: reciprocating and centrifugal. Most auxiliary pumps are centrifugal pumps, and the mud pumps are reciprocating.

Reciprocating Pumps

A *reciprocating pump* is a *positive displacement*, or plunger, *pump*. These names describe the way the pumps work. A *piston*, or *plunger*, reciprocates (moves back and forth) inside a cylinder, which is actually a replaceable liner. The reciprocating movement of the piston displaces (moves) fluid (fig. 43). Each movement of the piston in one direction is one *stroke*.

Most reciprocating mud pumps are either single-acting triplex pumps or double-acting duplex pumps. Today, most rigs use single-acting triplex pumps, although a few still use double-acting duplexes. A single-acting triplex pump is a pump with three pistons moving inside three liners (cylinders). A double-acting duplex pump is a pump with two pistons moving inside two liners.

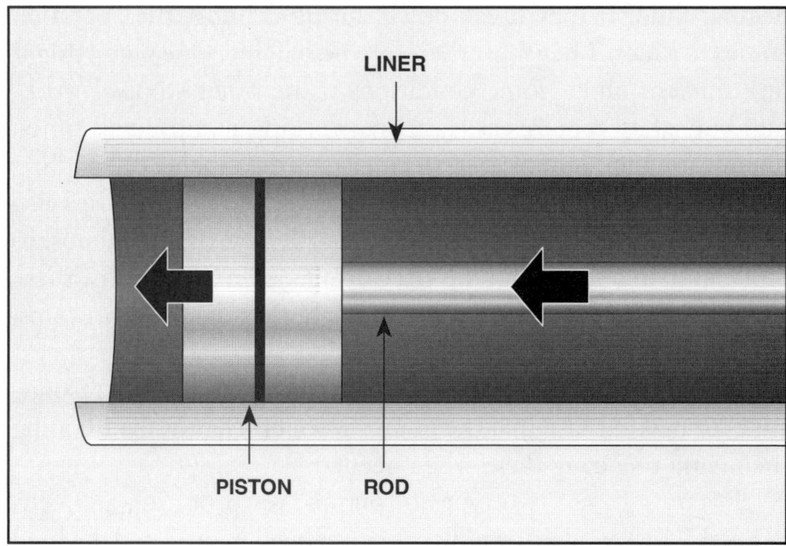

Figure 43. In a reciprocating pump, a piston moves back and forth in a liner.

In a single-acting pump, the pistons move back and take in mud through open intake valves. The pistons then move forward and push the mud out through open discharge valves (fig. 44a). Triplex pumps are single acting because they move mud out of the pump only in one direction.

In a double-acting pump, the pistons move back and take in mud through open intake valves. At the same time, on the other side of the same pistons, the pistons move mud out of the pump through open discharge valves. Then, when the pistons start back in the other direction, they pull in mud on one side and discharge mud on the other side (fig. 44b).

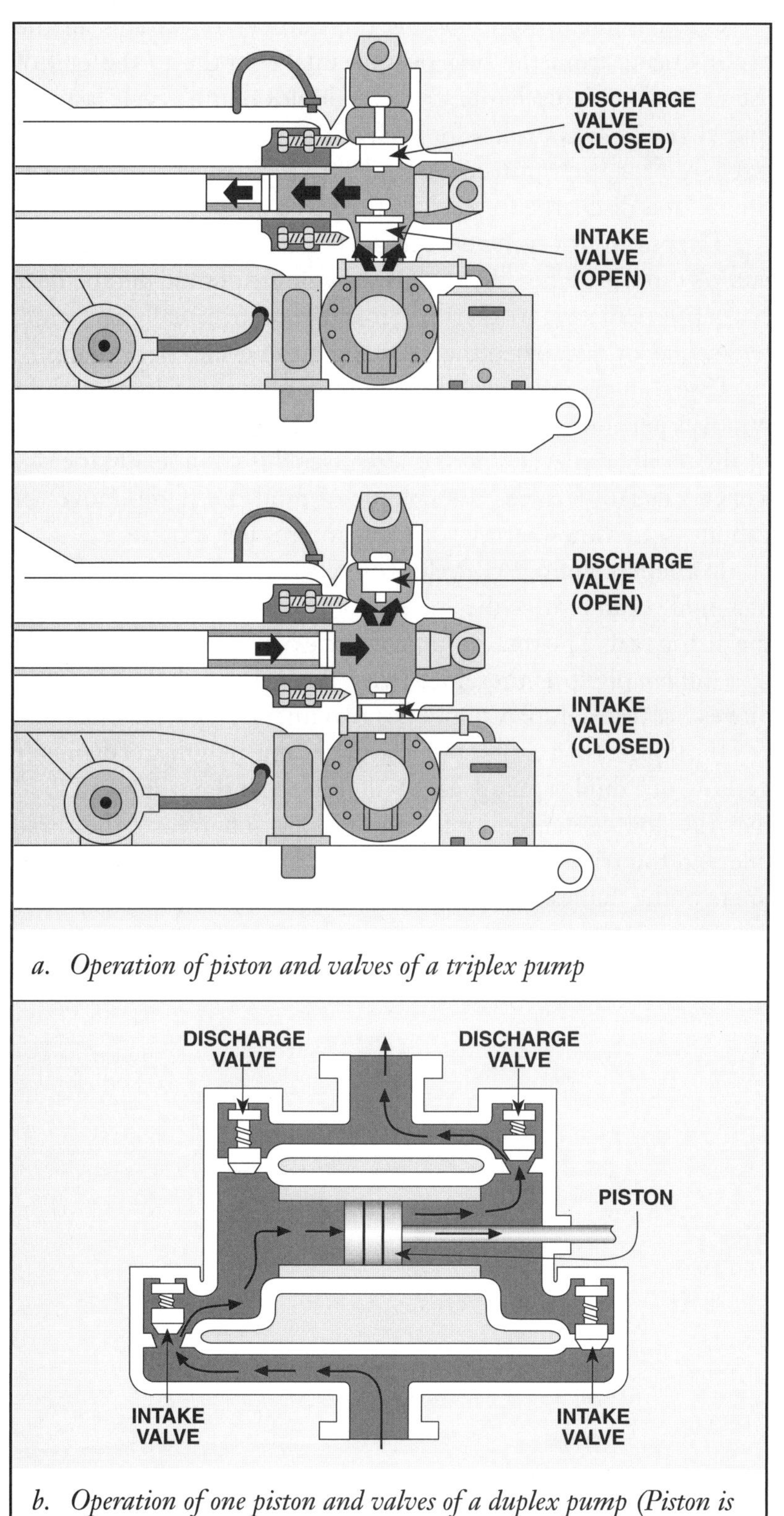

a. Operation of piston and valves of a triplex pump

b. Operation of one piston and valves of a duplex pump (Piston is moving from left to right.)

Figure 44.

One complete pump *cycle* is the number of strokes all the pistons make from the first piston's intake stroke to the end of the same piston's discharge stroke. A duplex pump's cycle lasts for four strokes—two strokes for each of the two pistons (intake and discharge for each piston). A triplex pump's cycle is six strokes (intake and discharge for each of the three pistons).

The number of cycles per minute is half of what determines the pump's capacity. The other half is the volume of mud pumped per cycle. (Note that rig workers and manufacturers usually say *strokes per minute*, or *spm*, when they actually mean cycles per minute.)

One way manufacturers rate a pump is by the volume of fluid it moves per minute. Multiplying the volume per complete cycle by the number of cycles per minute gives this volume. On the rig, workers usually express the volume of mud the pump moves in gallons (sometimes barrels) or cubic metres per minute.

In a duplex pump, each cycle of one piston results in the discharge of a mud volume twice the volume of the liner, less the volume of the piston rod. The total volume for one complete pumping cycle of a duplex pump is therefore twice the volume that one piston moves because a duplex pump has two liners.

If the liners of a triplex pump are the same length and diameter as those of a duplex pump, each will move half the volume of fluid that the double-acting duplex pump does. Therefore, the three liners in the triplex pump cannot move as much fluid as the two liners of the duplex pump. For this reason, triplex pumps operate at higher speeds to pump the same volume of mud.

Configuration of a Triplex Mud Pump

The *power end* of a reciprocating pump is the end connected to the prime mover, which drives the working parts. The other end of the pump, the *fluid end*, takes in mud and moves it out (fig. 45).

Figure 45. Power and fluid ends of a triplex pump (Courtesy of Nabors Drilling USA)

Power End

At the power end, rig-up crew members connect the mud pumps either to a compound, a series of gears and shafts that transfer power from the diesel prime movers, or to an electric motor. In the case of electric-motor driven pumps, the prime movers power electric generators, which supply electricity to the motors. On rigs that use a compound, crew members install power bands (large V-belts) between the compound and the pump's power end (fig. 46a). The power bands drive a pinion shaft inside the pump. The pinion shaft turns a small gear—the *pinion gear*. On electric-drive rigs, a chain and sprocket drive from the electric motor turns the pinion gear (fig. 46b).

Figure 47 shows the essential parts of the power end of a mud pump. The pinion gear drives a larger gear—the *bull gear* or *herringbone gear*, which is attached to the crankshaft. The crankshaft turns to give a back-and-forth motion to the *connecting rods*.

Figure 46a. Power bands (V-belts) from the rig's compound drive this triplex pump. (Courtesy of Nabors Drilling USA)

Figure 46b. *Electric motors turn a chain drive under the steel guard. (Courtesy of Nabors Drilling USA)*

These are linked to the *crossheads*, blocks that reciprocate in a straight line. *Pony rods*, or *crosshead extension rods*, connect the crossheads to the piston rods. Each piston rod is a metal shaft that attaches to the piston in the fluid end of the pump.

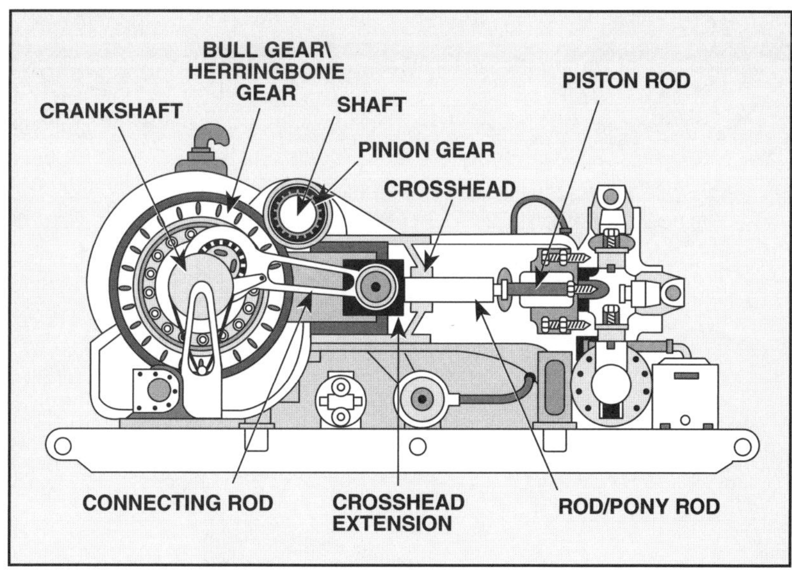

Figure 47. *Essential parts of the power and fluid ends of a mud pump*

Some pumps feature a chain and sprocket reduction drive instead of gears. Others use eccentric cams to operate the connecting rods, instead of a crankshaft. Other internal details of the power ends of these pumps, however, are the same as those for mud pumps with gears.

Lubricating System. All these moving metal parts require lubrication to minimize friction and make them last as long as possible. Some pumps use the rotation of the crankshaft to pick up lubricant, or gear oil, from a sump and distribute it to the bearings and crossheads. Large pumps have a small electric lube pump that splashes oil onto the metal parts when the mud pump is operating at a low revolutions per minute (rpm). To help keep the oil clean, magnets in various locations in the power end collect iron particles in the lubricant that may have worn from the metal parts.

Wipers on the pony rods are an important barrier between the power end and the fluid end (fig. 48). They keep the lubricant away from the fluid end, and drilling mud out of the power end. A grease fitting allows the crew to pump grease between the seals to form an additional barrier against mud entering the power end.

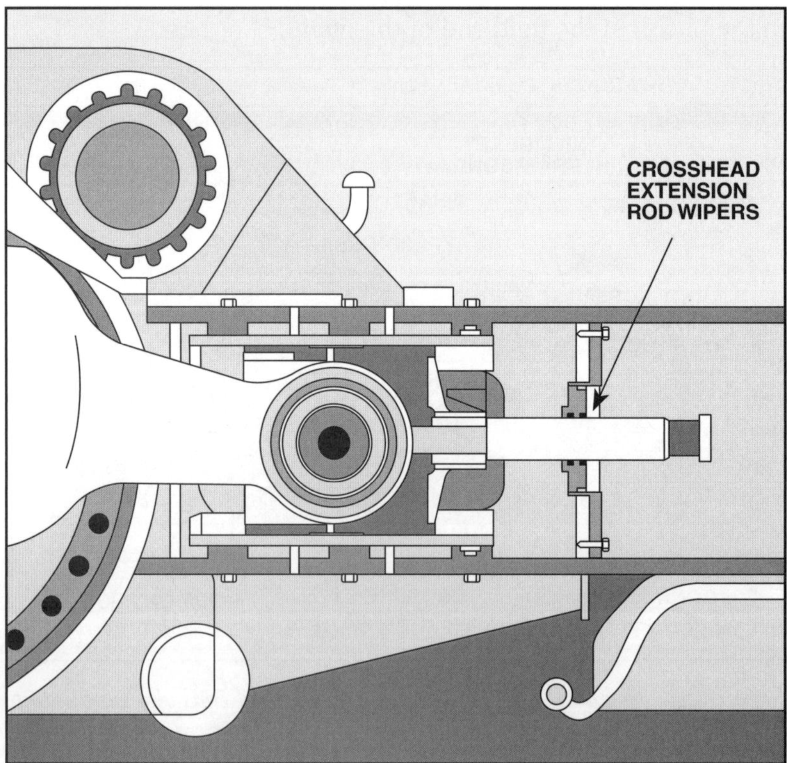

Figure 48. Pony rod wipers

Fluid End

The *fluid-end body* is made up of suction and discharge sections with machined cylinders, suction and discharge passageways, and openings (pots) to accept the valves (fig. 49). The crew can replace each suction and discharge section by removing nuts or screws from the studs that hold the sections together.

Inside the fluid-end body are the fluid end parts: the pistons on their rods, the liners, liner packings, and valves. Refer to *API Spec. 7* for standard dimensions and markings for fluid-end parts.

Figure 49. Part of a triplex pump's fluid end, showing valve pots

Pistons and Liners. The piston (fig. 50) is made up of two natural rubber or oil-resistant synthetic rubber cylinders wrapped around a metal core, the *piston body*. A flange on the piston body has a *wear groove* and *bypasses* cut into its circumference. Measuring the depth of the wear groove and comparing it to the original depth indicates how much the piston body has worn. The bypasses are grooves for relieving excess pressure.

The piston fits snugly inside a replaceable steel liner (fig. 51). The liner, in turn, fits inside the pump. The type of steel determines the liner's resistance to corrosion and abrasion. The pressure seal between piston and liner must be effective, although the piston must still be able to move freely inside the liner.

A spray system cools the pistons and liners. This system consists of a small pump (fig. 52), a manifold with nozzles that continuously spray the coolant, and a sump. The coolant may be water or soluble oil. Clean, cool water is the most economical.

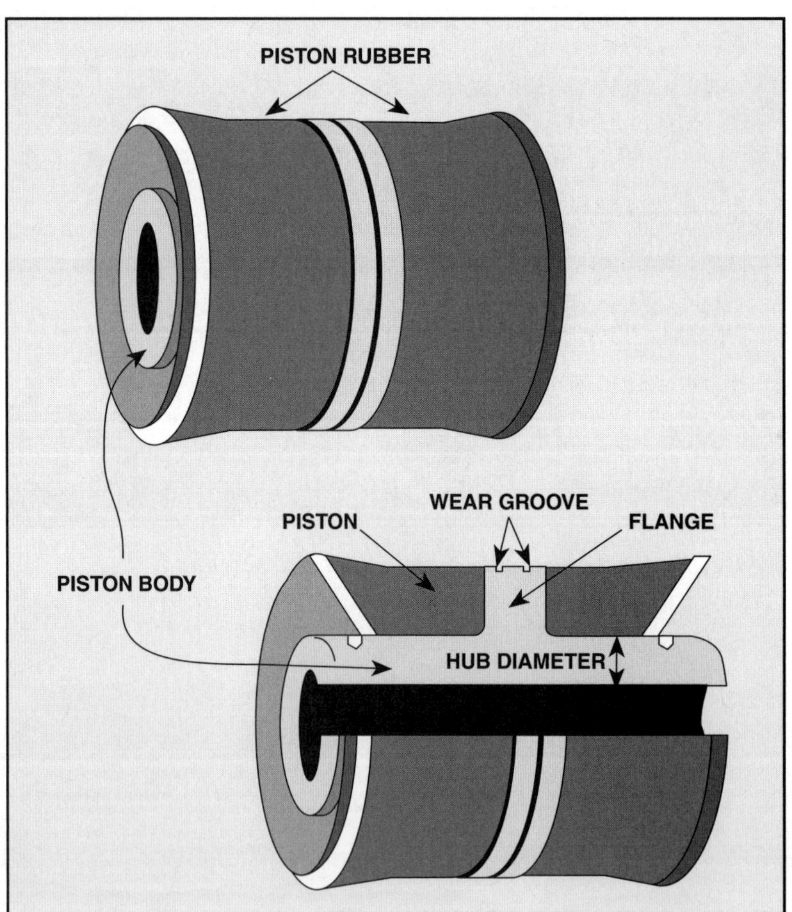

Figure 50. Pump piston

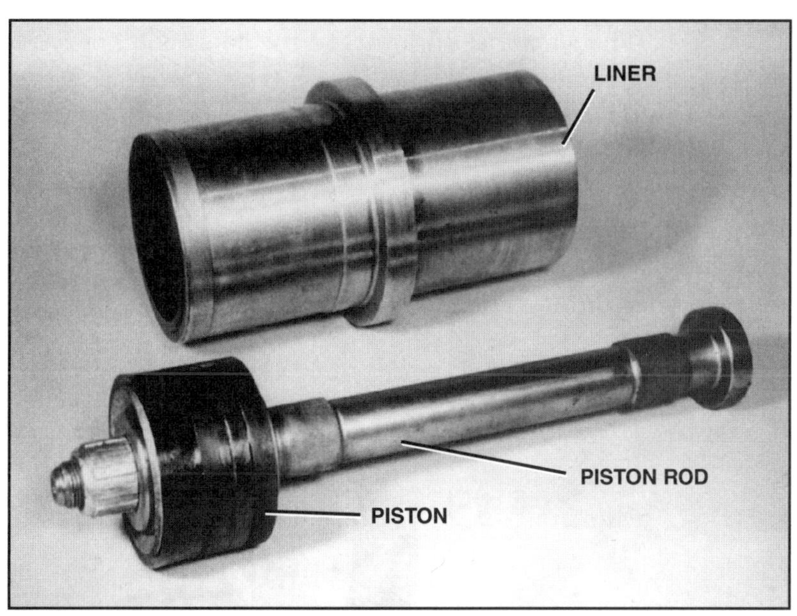

Figure 51. Liner, rod, and piston for a triplex pump

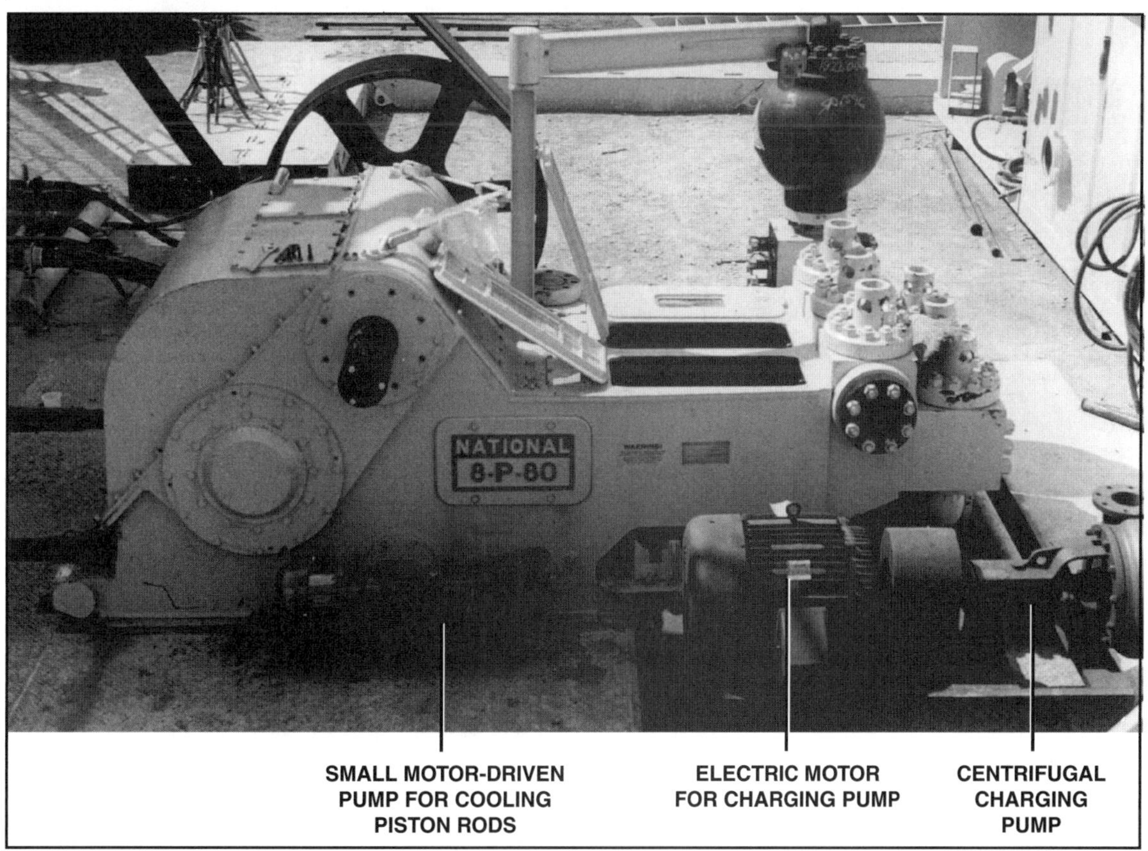

Figure 52. Motor-driven pump for cooling piston rods and a supercharging pump

Piston Rods. The *piston rods* (fig. 53) are the connection between the power end and the fluid end. Each rod has a narrow section on one end that fits inside a hole in the piston body. A shoulder fits against the shoulder of the piston body and a nut on the end tightens the two. The other end of the piston rod is threaded and has a shoulder that tightens against the pony rod.

Crew members can remove a cover on the pump's body to get to the rods when they need to repair or replace them or the pistons.

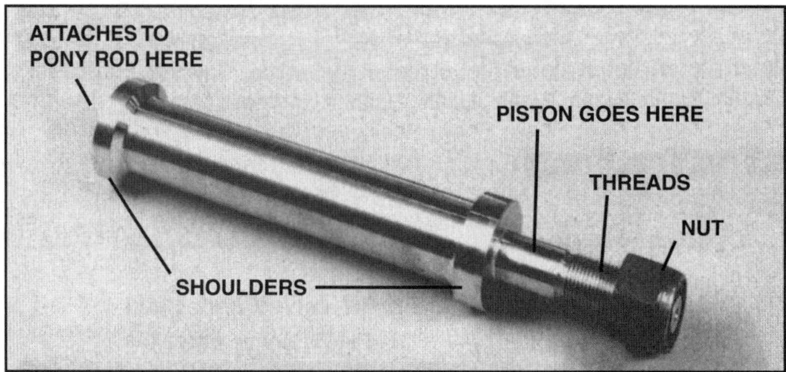

Figure 53. Piston rod (Courtesy of National Oilwell, Mission-Fluid King)

Liner Packing. Liner packing makes a pressure seal between the liner and the interior part of the pump that holds the liner. It consists of several metal and plastic rings stacked one against the other (fig. 54). The rings, at left in the figure, fit together to form the packing on the right. The center ring, the *lantern ring*, has a notch that lines up with a grease fitting. When lubricant is injected into the fitting, the lantern ring distributes it to the whole packing.

Figure 54. Liner packing (Courtesy of National Oilwell, Mission-Fluid King)

Some pump manufacturers drill a small hole, the *weephole*, *tattletale hole*, or *telltale hole*, in the pump's body that lines up with the packing. The hole leaks fluid when the liner packer is failing, as long as the hole is not plugged. When rig personnel notice any leakage, they know it is time to change the packing. Be sure to keep the telltale hole clean, if your pumps have them.

Valves. Each cylinder of a single-acting pump has two valves. (Figure 55 shows two different types of valve and valve seat.) The pump piston draws mud in through the *suction valve* and pushes it out through the *discharge valve*. The valves move up and down inside *valve seats*. Intake or discharge pressure opens the valves. A heavy-duty spring keeps the valves closed in the absence of intake or discharge pressure. The seats fit into *valve pots* machined into the fluid-end body (see fig. 49). The valves must be free to move vertically in the seats, but still form a seal when they are in the closed position. Replaceable rubber or polyurethane inserts make the seal.

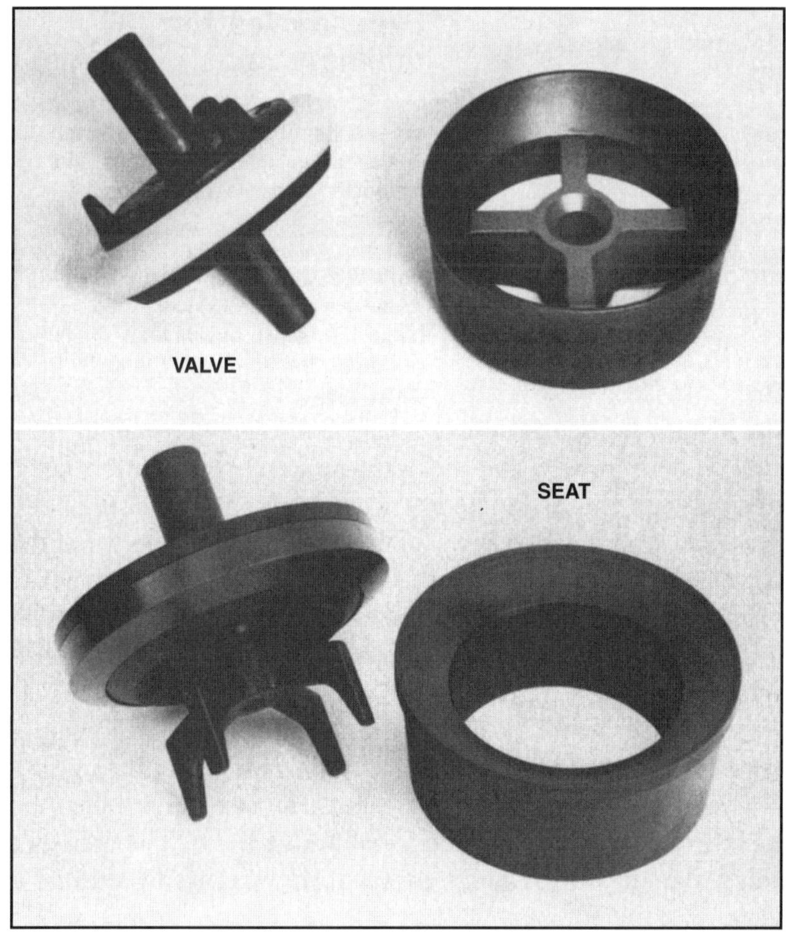

Figure 55. Valves and valve seats (Courtesy of National Oilwell, Mission-Fluid King)

Configuration of a Duplex Mud Pump

A duplex pump (fig. 56) is more complex than a triplex because of the double-acting design. The power end, pistons, rods, and liners have similar constructions but are larger and heavier. Each cylinder, however, has a special packing to seal off the piston rod because fluid flows on both forward and backward strokes. A duplex pump has four intake and four discharge valves, while a triplex pump has three intake and three discharge valves.

Figure 56. Duplex, double-acting pump

Rods and Rod Packing

Because the piston of a double-acting pump encounters fluid pressure on both forward and backward strokes, manufacturers taper the rod where the piston body fits onto the rod. The piston body has a matching tapered bore. Thus when crew members install a new piston body, they fit the body's taper into the rod's taper and firmly mate the two, usually with a special hammer. The taper-to-taper fit creates a firm attachment that a straight bore cannot achieve.

A cross section of the fluid end of a duplex pump (fig. 57) shows that the piston rod strokes through a *stuffing box*. Inside the stuffing box is a *gland* that compresses the synthetic or natural rubber *rod packing* that wraps around the piston rod (fig. 58). The crew can adjust the amount of compression of the packing by turning a gland-packing nut.

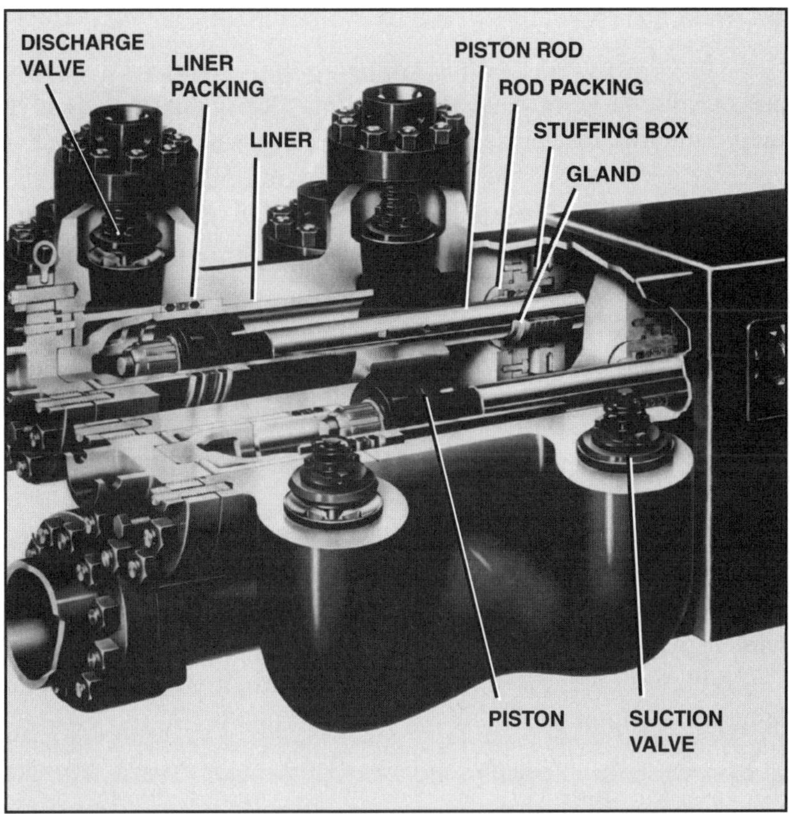

Figure 57. Fluid end of a duplex pump (Courtesy of National Oilwell, Mission-Fluid King)

The rod packing seals the opening between the rod and stuffing box, which keeps out drilling fluid, including particles in it that could score the rods. Rod packing also lubricates the piston rods to decrease wear to the stuffing box and piston rods.

Figure 58. Piston rod packing (Courtesy of National Oilwell, Mission-Fluid King)

Valves

At each end of each cylinder in a duplex pump are one suction valve and one discharge valve (see fig. 57), for a total of eight valves: four intake and four discharge. The movement of the piston causes the valves to act in pairs—the intake valve on one end opens while the discharge valve on the same end closes, and so on. They fit inside valve seats and valve pots like those on a triplex pump.

Pumping Output of Reciprocating Pumps

The volume of mud pumped by a reciprocating pump depends on several factors: the speed of operation, the area of the piston, the diameter of the liner, and the stroke length. The volume varies directly with the speed of operation in strokes per minute (spm). In other words, the faster the pump pumps, the more volume it pumps. Pumps with a short stroke can usually move at higher speeds than longer-stroke units. When the speed is constant, a pump with a large-diameter liner moves more mud than a pump with a small-diameter liner.

Table 6 shows how the volume varies with liner size and stroke length in a triplex pump.

Table 6
Nominal Pumping Rates of Triplex Single-Acting Pumps at 130 spm
(gallons per minute; m^3/min)

Stroke	Liner Sizes (in.; mm)				
(in.; mm)	7 (177.8)	6½ (165.1)	6 (152.4)	5½ (139.7)	5 (127)
8 (203.2)	- (-)	- (-)	382 (1.446)	321 (1.215)	265 (1.003)
9 (228.6)	- (-)	- (-)	429 (1.624)	360 (1.363)	297 (1.124)
11 (279.4)	715 (2.706)	616 (2.332)	525 (1.987)	441 (1.669)	- (-)

Volumetric Efficiency

Volumetric efficiency is the volume the mud pump actually pumps compared to the volume that calculations say it should. It is a measure of the overall work produced by the pump, as well as of its effectiveness. For example, if the theoretical volume that the pump can move is 800 gallons per minute (3 cubic metres per minute), and it actually moves only 680 gallons per minute (2.5 cubic metres per minute), the volumetric efficiency of the pump is 85 percent (680 divided by 800; or 2.5 divided by 3).

Very few duplex pumps achieve much more than 90 percent volumetric efficiency because some liquid always leaks between the piston and liner on each stroke. Triplex pumps, however, are closer to 100 percent efficient. Mud cut with air or gas is the most frequent cause of reduced volumetric efficiency. Entrained gas in the form of tiny bubbles can be a serious problem. A mud containing 15 percent gas may cause a reduction in volumetric efficiency of 15 percent because the gas takes up 15 percent of the space in the liners that mud could fill.

Other reasons for a drop in efficiency are poor suction and failure of parts. Faulty or failing parts can result in leakage of the mud between piston and liner, liner and fluid-end body, or valve and valve seat. Such leakage causes *fluid cutting* or *washouts*—that is, it abrades the metal parts. Fluid cutting is usually severe before anyone notices a pressure reduction or loss of volumetric efficiency.

Controlling Pump Output

Manufacturers rate pumps by the amount of input power (in horsepower or kilowatts) they can use, their maximum speed in strokes per minute, the maximum amount of pressure (in pounds per square inch or kilopascals) they can produce, and the volume of mud (in gallons or cubic metres per minute) they can pump at 100 percent volumetric efficiency. How much horsepower a pump uses varies with volume times pressure. A reciprocating pump pumps a constant volume; therefore, the higher the pressure, the higher the power requirement.

Reciprocating pumps require the maximum rated power input to operate at full rated speed and pressure. Usually, however, real-life drilling requires that the driller operates the pump at something less than top speed and pressure. The driller can vary the pump speed and pressure by changing the input power or changing the size of the liners. A larger liner holds a larger amount of mud on each stroke, so it pumps more mud than a smaller one at the same speed. But it pumps it at a lower pressure. Similarly, small liners pump at high pressure and low volume. The size of the liner limits the discharge pressure, regardless of speed. So if the engine overloads or the discharge pressure is higher than needed, the crew will install a smaller liner.

Each brand and size of pump usually has a nameplate that tells rig personnel how much pressure various sizes of liners will produce. Different pumps may have different pressure ratings for the same liner diameter, depending on their design and input power rating.

Suction

Ensuring that a mud pump has a flooded suction is one of the most important things crew members can do to ensure that the pump is running at maximum efficiency. *Flooded suction* means that the fluid end is always full of mud while it is pumping. If the pump's fluid end is not completely full of mud, air can enter the intake line. The pump then draws in mud with air in it, which reduces volumetric efficiency. In short, the pump is pumping not only mud, but also air.

To guarantee a flooded suction: (1) completely fill (prime) the pump's fluid end with mud before starting it, and (2) make sure that only mud (not air) enters the pump while it is working. Submerging the pump's intake, or suction, line completely in the suction tank prevents air from entering the line at the tank. But a completely submerged line alone does not ensure flooded suction. At high speeds, the pump pistons can move faster than heavy muds. The fast moving pistons "outrun" the heavy mud and leave a gap of air between the piston and the mud. To overcome this problem, crew members can install a supercharging pump.

Supercharging Pumps

A *supercharging pump* (see fig. 52) keeps the suction manifold and liners full of mud at any speed. The supercharging pump picks up mud from an active tank and sends it to the mud pump's suction line. It ensures that the mud pump's intake, or suction, is full of mud at all times. A supercharging pump therefore maintains a flooded suction. A flooded suction ensures that mud is always in contact with the pistons on the intake stroke.

Supercharging pumps are usually centrifugal pumps. Rig owners should make sure that the centrifugal pumps are large enough to move as much mud as the main mud pump when the main pump is running at maximum speed with its largest liner size. Either a belt from the pinion shaft of the reciprocating mud pump or an independent power source drives the supercharging pump.

Pressure Surges

Mud comes out of a reciprocating pump in a surge-pause, surge-pause manner (fig. 59). In a double-acting duplex pump, one piston moves forward, pushing mud ahead of it to the front of the pump, pauses, then starts back, pushing mud in the other direction, all the way to the back end of its stroke. At the same time, the second piston is making the same back-and-forth pulsing stroke, except always in the opposite direction to the first piston.

In a single-acting triplex pump, one of the three pistons strokes when the crankshaft moves one-third of the way around, each 120 degrees of rotation. So each full revolution of the crank produces three surges in the mud flow, one from each cylinder. In both types of pumps, the piston moves faster than the viscous mud can flow, so a small empty space forms next to the piston on each stroke. As the piston comes to the end of its stroke and starts back, the mud hits its face. The piston must then overcome the force of the mud flowing forward as it starts backward. The result is that the piston moves unevenly, which causes a pulsating flow and fluctuations in pressure within the mud pump, in the surface piping system, and in the drill string. Because mud does not compress and the heavy piping is quite rigid, the system cannot easily absorb the changing pressure.

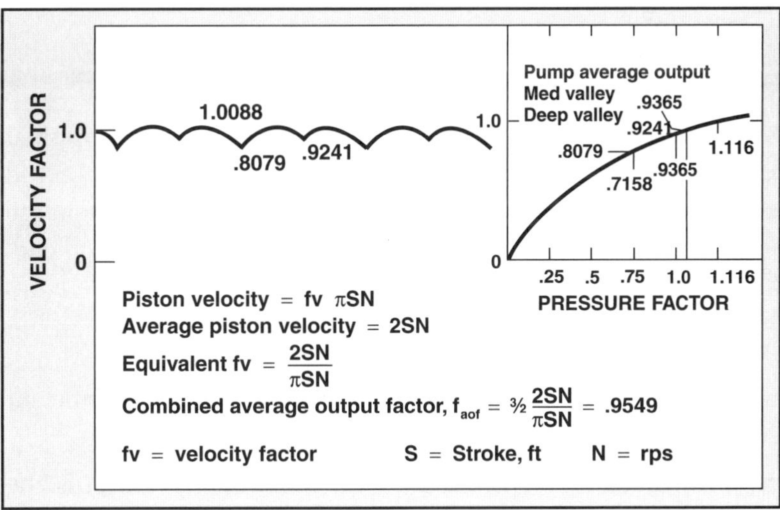

Figure 59. The bumpy line to the left indicates surges in pump delivery.

A supercharging pump can smooth the discharge flow, but pressure surges are normal in reciprocating pumps. Eventually the varying stress of pulsing causes surface pipe to fail, especially at the connections. It also creates vibrations that can cause connecting hoses to swell, vibrate, or whip. Pulsating flow worsens another problem, *cavitation*, in which the moving mud forms a void inside a sharp curve of piping (fig. 60). Cavitation causes erosion and abrasion, vibrations, and an audible crackling sound. For this reason, all pipes that carry drilling mud should have long, generous curves.

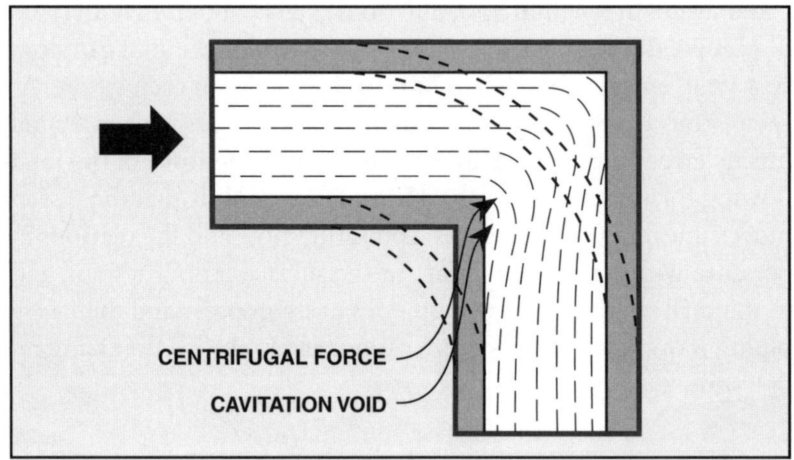

Figure 60. Cavitation occurs where fluid passes through sharp ells under high pressure.

Factors That Increase Pressure Surges

A number of conditions increase pressure surges. One is a normal part of operation in a double-acting duplex pump. The piston rod takes up space in the rod end of the cylinder. For this reason, the head end of the cylinder can contain more fluid than the rod end. The strokes that discharge mud from the head end move more fluid and produce greater pressure surges.

Two abnormal conditions that can increase these pressure surges even more in both types of reciprocating pumps are insufficient suction head and fluid knock.

Insufficient Suction Head. Suction head (or *suction pressure*) is the pressure of the fluid entering the intake valves of the pump. Because a mud pump may pull hundreds of gallons or litres of mud per minute, the suction hookup is crucial. It must provide an uninterrupted flow of mud into the pump. When the fluid does not completely fill the cylinders or fills them unevenly, excessive pressure surges occur. Insufficient suction head can also produce poor volumetric efficiency and sometimes fluid knock. (Fluid knock is covered in detail later.)

One cause of insufficient suction head is fluid friction in a long suction line. Another is any suction arrangement that requires the pump's suction to lift mud upward, against gravity. For this reason, rig owners should always install the mud tanks higher than the pump's intake (suction). A supercharging pump also prevents insufficient suction head.

Fluid Knock. Fluid knock, also called *hydraulic hammer* or *water hammer*, is closely related to insufficient suction head. Fluid knock results when the incoming mud cannot keep up with the piston as it speeds up during the stroke. As the piston reaches the end of the stroke, the fluid suddenly catches up with the piston and produces an audible knock as it hits the piston (fig. 61). The piston rod carries this shock back to the power, which puts added stress on the power-end parts. Fluid knock also causes the suction valve to strike its seat hard, which can cause premature wear on the valve seals.

Fluid knock also causes metal fatigue in the pump parts and the discharge line. Keeping the suction line short and straight reduces the chance of fluid knock. If the problem continues, the contractor may add a supercharging pump. Also, because triplex pumps usually move at relatively high speeds, supercharging pumps on triplexes are essential to prevent fluid knock.

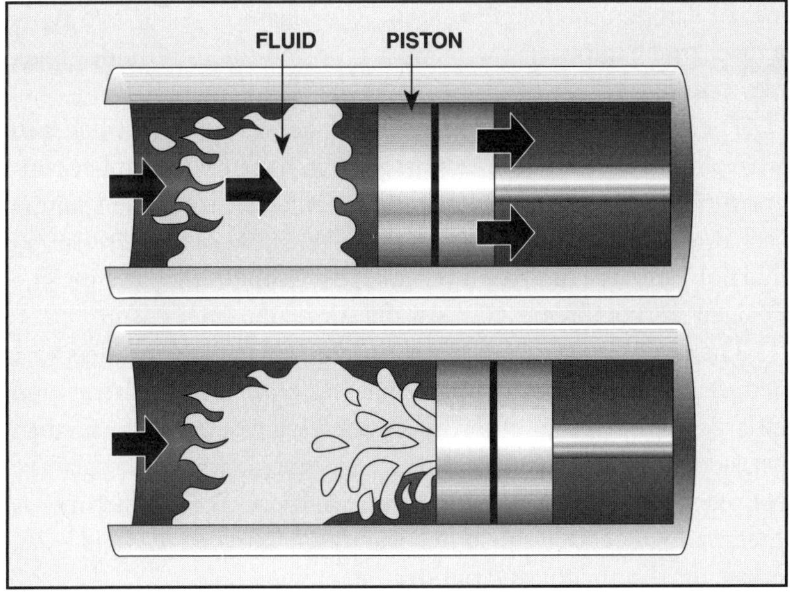

Figure 61. At top, the piston moves to right faster than mud can travel, leaving a gap between the mud and the piston. At bottom, the piston momentarily comes to a stop, allowing mud to catch up to the piston. The mud hitting the piston causes a knocking sound.

Figure 62. Suction and discharge dampeners (Courtesy of Nabors Drilling USA)

Pulsation Dampeners

Dampeners are devices that absorb pressure variations. The operator usually installs a dampener on the suction line, to dampen surges resulting from the pump's suction (fig. 62), and on the discharge line, to dampen the outgoing flow. Manufacturers offer two types of discharge dampener: the bladder type and the nonbladder type. *Suction dampeners* are of the bladder type only.

The bladder type of discharge dampener is a sphere divided into two chambers by a flexible diaphragm (fig. 63a). One chamber contains nitrogen gas. The other chamber is connected to the mud pump's discharge line. On each discharge stroke, some of the discharged mud flows into the mud chamber and pushes on the diaphragm. The nitrogen gas compresses to resist the surge of mud.

The nonbladder type is simply a large sphere, about 4 feet (1.2 metres) in diameter, installed in the line (fig. 63b). It fills with mud and the large volume of mud absorbs the pulses. It is easier to maintain than the bladder-type dampener because it has no moving parts. The bladder type requires that the diaphragm be replaced periodically, which means that crew members have to loosen and unscrew several nuts and bolts, replace the diaphragm, and tighten the nuts and bolts.

Dampeners come in different sizes to match the pump size. As described above, normal pressure surges in a double-acting duplex pump are greater than in a single-acting triplex pump, so duplex pumps need bigger dampeners to accomplish the same dampening effect.

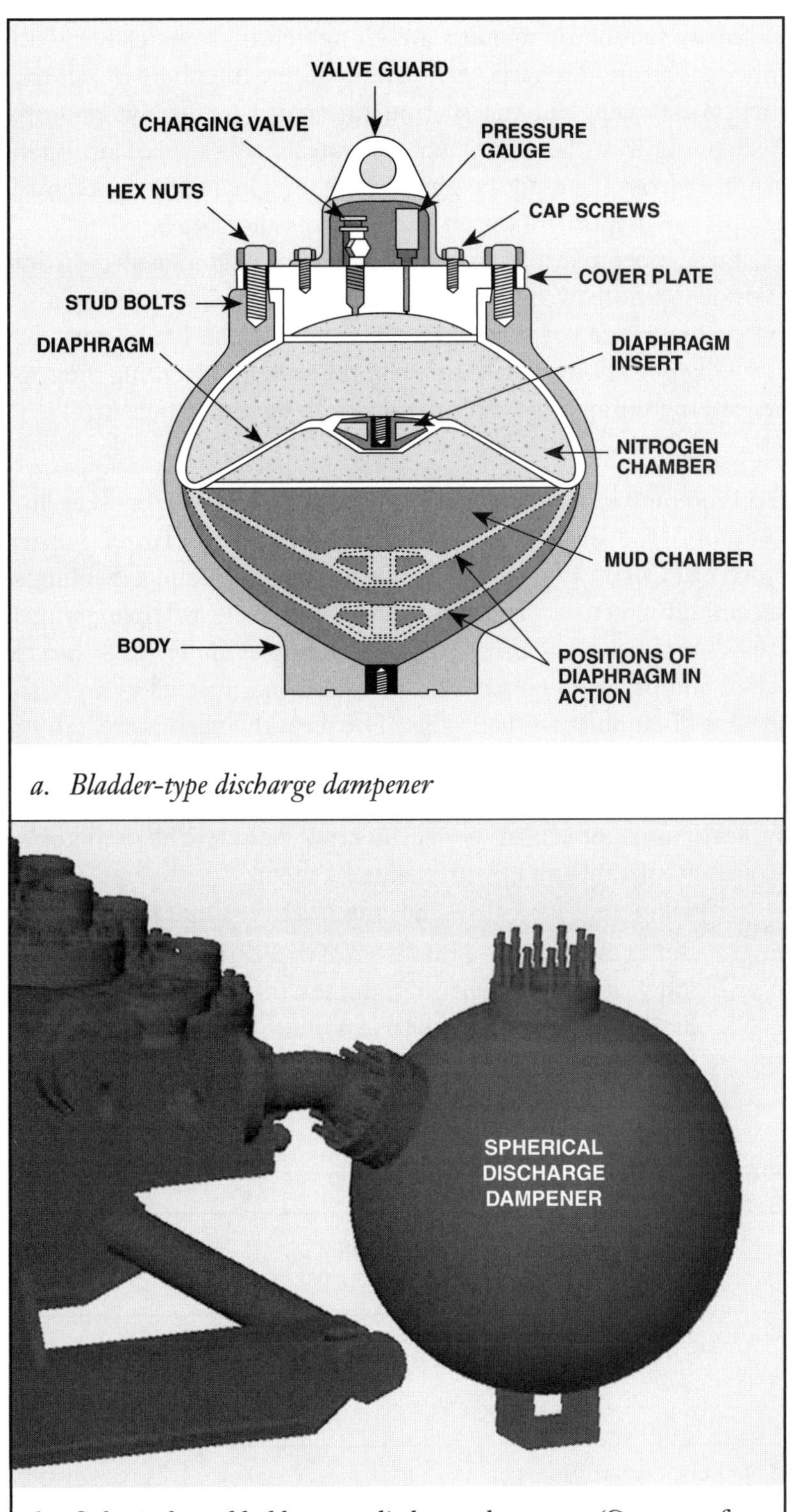

a. Bladder-type discharge dampener

b. Spherical, nonbladder-type discharge dampener (Courtesy of White Rock Engineering and National-Oilwell)

Figure 63.

Most suction dampeners are elongated in shape rather than spherical as are discharge dampeners. Crew members install the suction dampener in the suction line close to the mud pump. A diaphragm in the dampener separates a pressurized nitrogen chamber from the mud the pump takes in. The nitrogen pressure against the diaphragm offsets suction pressure surges.

Sized, operated, and maintained correctly, dampeners even out hydraulic power output and increase pump efficiency. They also cut maintenance costs because they increase the life of parts by reducing pulsation and vibrations in the system. Installing them as near to the pump as possible produces the greatest benefit.

Comparison of Triplex and Duplex Pumps

Drilling contractors used duplex double-acting pumps almost exclusively until the early 1960s. As duplex pumps became larger—up to 1,500 hp (1,050 kW)—weight became a real problem. The pumps became difficult to transport from one job to the next. Triplex pumps are lighter because their three single-acting cylinders need heavy valves and other metal parts only on one end, instead of on both ends, as do double-acting pumps. The parts themselves are lighter as well (table 7). Today, duplex pumps are a special-order item from manufacturers; virtually all new pumps are triplex. However, older duplex pumps remain in service, so crew members need to know how they operate and how to maintain them.

Triplex pumps have many advantages over duplex pumps:

1. Better efficiency—The design of triplex pumps makes them much more efficient than duplex pumps. Duplex pumps have many more places to leak fluid; the seals around the

Table 7
Comparative Weights of Pump Liners

Liner Size inches (millimetres)	Weight of Liner pounds (kilograms)			
	Duplex (500 hp; 375 kW)	Triplex (500 hp; 375 kW)	Duplex (1,000 hp; 750 kW)	Triplex (1,000 hp; 750 kW)
4 (101.6)	338 (153)	98 (44)	- (-)	119 (54)
5 (127)	313 (142)	81 (37)	466 (211)	98 (44)
6 (152.4)	259 (117)	55 (25)	394 (179)	67 (30)

rods create more friction; and the triplex pump's high speeds (up to 175 spm, compared to a duplex's maximum of 70 spm) create more momentum in the fluid.

2. Better power-to-weight ratio—Triplex pumps are 40 to 50 percent lighter than duplex pumps rated at the same horsepower.
3. Higher pressure—Triplex pumps can produce a discharge pressure up to 6,000 psi (41,370 kPa), compared to the maximum duplex pressure of about 2,000 psi (13,790 kPa).
4. Smoother discharge flow—A triplex pump produces a smoother flow than a duplex pump because it is single-acting.
5. Ease of repair—Fluid-end parts are smaller and more accessible in the triplex pump. Liners and pistons can be replaced in less than half the time needed for a duplex; and a triplex rod, for example, weighs only some 30 pounds (13.5 kilograms), compared to a duplex rod's 200 pounds (90 kilograms).
6. Need less space—Triplex pumps take up less space, which is especially desirable offshore where space is at a premium.
7. Less expensive parts—Replacement parts for a triplex pump cost less than half those for a duplex because they are smaller and lighter.

To summarize—

- Mud pumps are reciprocating pumps. They may have two cylinders (duplex) or three (triplex). Triplex mud pumps are single-acting, and duplex are double-acting.
- Gas-cut mud, poor suction, and failure of parts cause the volumetric efficiency of a mud pump to drop.
- Supercharging pumps keep the mud pump primed by maintaining flooded suction in the fluid end.
- Because the reciprocating action of the pistons moves mud in a pulsing manner, mud pumps use pulsation dampeners on the both the suction and discharge lines.
- Triplex pumps have replaced duplex pumps on most rigs because of their many advantages, including lighter weight, smaller size, and ease of maintenance.

Operating Reciprocating Pumps Efficiently

To get the best work from either a duplex or a triplex pump, maintaining good suction and discharge flow are the most important things a crew can do.

Suction

Poor suction wears out pump parts quickly, and replacing failed parts is expensive. To maintain good suction, the suction line from the mud tank to the pump should be large, short, and straight (fig. 64). Changes in size and direction of the line cause pressure losses. Crew members should fully open any valves in the suction line. Further, rig owners should only install valves that, when open, are the same size or larger than the suction opening of the pump. The opening between the suction line and the suction dampener should be as large as practical, as any restriction reduces the effectiveness of the dampener.

Use a supercharging pump that provides the correct discharge pressure. The pressure on the mud provided by the supercharging pump should be neither too high nor too low. A supercharging pressure that is too high at the mud pump suction may cause sluggish valve action because the valves have to open against pressure. On the other hand, pressure too low at the suction allows shock and fluid knock because there is not enough pressure to fill the suction line.

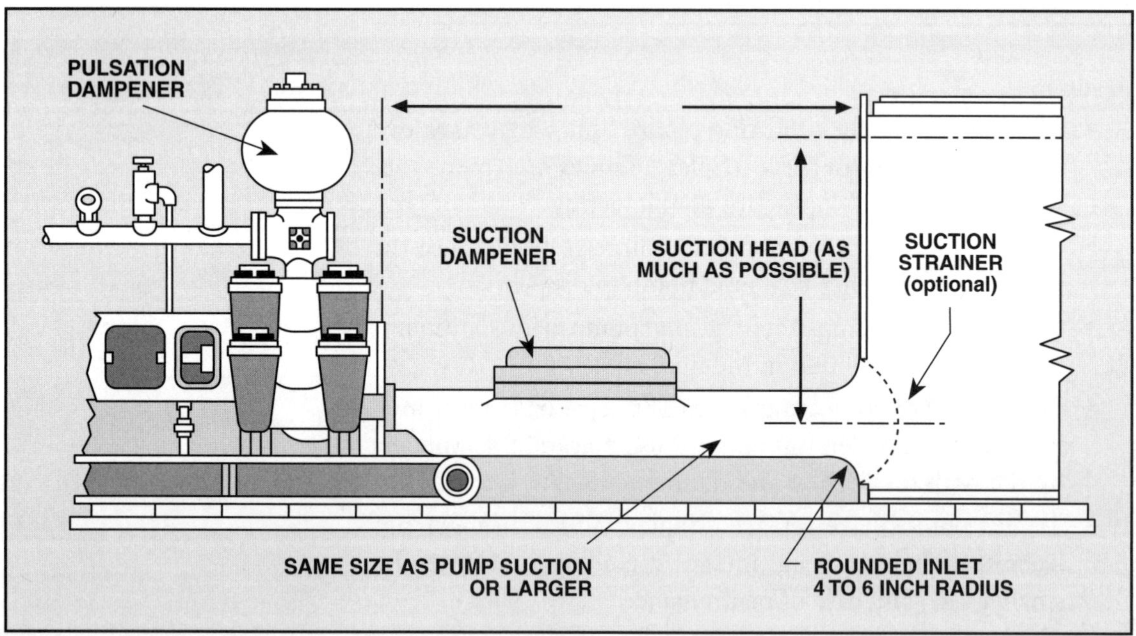

Figure 64. Suction line from suction tank to pump

Some rigs owners install a suction-line relief valve. They install it on top of the suction line between the suction dampener and the pump's intake opening. They mount the suction relief valve on top of the suction line so that gelled mud cannot clog it when the driller shuts down the pump. Installers normally adjust the valve to open (relieve) if pressure inside the suction line goes above 70 psi (500 kPa).

A suction relief valve protects the supercharging pump and the suction line dampener. (Recall that the supercharging pump puts mud into the suction line to keep the pump's suction flooded, or full of mud.) In the main mud pump, the pump pistons pull mud in through open suction valves, and push it out through open discharge valves. If both a suction valve and a discharge valve failed in a closed position on the same side of the pump, high back-pressure—a pressure surge—would occur. The mud that the stroking pump pistons move could not escape because both valves are closed. This surge would flow back into the suction line. If the surge were above 70 psi (500 kPa), the suction line relief valve would open. The open relief valve would keep the pressure surge from damaging the suction line dampener and the supercharging pump.

Discharge Flow

To keep the discharge flow continuous, crew members should rig up the discharge piping as simple as possible. Every elbow, tee, or bend, any valve that does not open fully, or any change in pipe size, may cause fluid knock and severe vibrations. Long, slowly angling bends are better than sharp right-angle turns. Anchor all sections of the line firmly to a rigid structure, and tie down the ends of high-pressure hoses securely so that they cannot whip free in case of a break. The crew should weld piping rather than screw it together whenever possible.

The discharge line often has a *pressure-relief valve* immediately next to the pump (fig. 65). This valve protects the pump when the discharge line or a bit nozzle becomes plugged. It opens automatically when the pressure in the line becomes higher than a certain preset value. If the valve opens too often, the preset value is too close to the operating pressure.

To ensure that the pressure-relief valve itself does not get plugged, the discharge line also contains a screen, or *strainer*, which screens out any large pieces of material. The pressure-relief valve must sit in front of the strainer, and the line must not have a shutoff valve between the pressure-relief valve and the fluid end. Such an arrangement would prevent the relief valve from protecting the pump.

If the discharge pressure rises too high, the pressure-relief valve opens, and the mud enters a relief bypass line. This line, which returns fluid to the mud tanks safely, should be short, preferably without bends, and rigidly anchored. Never return the relief bypass to the pump suction. To keep the bypass line open, do not allow it to stand full of mud in freezing weather or to become plugged with dried mud.

Figure 65. Pressure-relief valve on pump discharge line (Courtesy of Nabors Drilling USA)

Knocking

A mud pump is normally noisy when it runs. Crew members should, however, investigate any unusual knocking sound. Mechanical knocking, caused by a metal part banging on another part, is usually a sharp, high-pitched, metallic sound. Fluid knocking, on the other hand, is lower pitched (deeper in tone). Mechanical knocking can occur at all normal pump speeds.

If the noise is coming from the power end of the pump, it is almost invariably mechanical knocking. If the noise is coming from the fluid end, it may be either mechanical or fluid knocking. In this case, attempt to locate the source by using a wrench, welding rod, or hard hat as a stethoscope. This simple technique usually makes it possible to determine which cylinder is causing the noise, and sometimes to locate the exact valve pot. If the noise seems to be coming from one spot, it is usually mechanical. If it seems general and is hard to locate, it may be fluid knock. Triplex pumps are more prone to fluid knocking than duplex pumps because the pistons move so fast.

Table 8 offers suggestions about what to check when various trouble symptoms develop.

Table 8
Locating Sources of Pump Troubles

Symptom	Checks
Mechanical knocking occurs immediately after pump is put back into service following replacement of parts.	Determine whether pump has been properly primed. Check replaced parts for correct size, type, and installation. Check liners for tightness. Check rod nut, fit of rod in the pony rod, makeup of locknut. See that the packing is tight by checking the gland-packing nut. If knock persists, suspect a loose piston, which will cause a knock each time the piston changes direction. By holding the rod, you can feel the slamming of the piston.
Mechanical knocking develops in the fluid end.	Check pump for loss of prime. Redo steps listed previously, especially checking for locknut tightness or broken parts. Inspect valves for lost circulation or foreign materials. Make sure the piston rod has not become partially unscrewed, allowing rod to hit liner cage.
Mechanical knocking develops in the power end.	Consult pump manufacturer after checking crosshead guides for wear and pony rods and locknuts for tightness.
Fluid knocking develops.	Observe suction hose for excessive surging or whipping, indicating that the fluid knock may be caused by an improperly charged pulsation dampener. Observe rotary hose for excessive whipping, indicating that the fluid knock may be caused by an improperly charged pulsation dampener.

Table 8, cont.

Symptom	Checks
Pump knocks at high speeds only.	High-speed knocking often indicates incomplete filling of cylinder, caused by many possible conditions. 1. Check suction strainer for clogging. 2. Check mud pits for large deposits of sand and drilled solids. 3. Check suction hose for collapse or deterioration. 4. Check mud level; if it is too low, air will be drawn into the suction. 5. Check suction manifold lines to make sure the correct valves are fully open and not sucking air at the stem; check couplings and flanges also. 6. Check mud temperature. If it is too high, increase mud circulation or mud pit area. 7. Check suction valves for possible sticking. Lost circulation materials may interfere with valve lift.
Knocking persists after all checks have been made.	If suction conditions are poor and corrective measures cannot be made, use a centrifugal precharging pump.
Rotary hose vibrates excessively, although pump is not knocking.	A pulsation dampener may be needed. If one is being used, check precharge pressure and check for leaks in the dampener.
Rotary hose whips, and pump is knocking. causes all	Gas-cut mud is probably being pumped—a dangerous and expensive condition that must be corrected immediately. Severe gas cutting the rotary hose to whip but the suction hose merely to vibrate. Check suction lines for air leaks or an air pocket at a high point.
Pressure drops abnormally, although no knocking occurs.	Check each valve pot by ear for washing sounds. Also check cylinder. If the sound seems uniform throughout piston travel, check piston rubber for leaks. If no washing sounds are heard, check the discharge strainer for clogging and the drill string for leaks.
Discharge pressure is excessive or fluctuating abnormally.	Check both cylinders immediately for washing sounds, indicating piston leakage. Prolonged leakage will damage the liner. If pistons are all right, check valve pots for washing sounds.

Priming

Priming the pump (filling it with fluid before starting it up) is crucial, because just a few strokes of a piston in a dry liner may ruin the piston rubber.

Always prime the pump after replacing parts. Flooded suction from a mud tank helps maintain the prime in a pump. A supercharging pump can quickly prime the slush pump before it is started, as long as no back-pressure is present on the discharge side of the pump, and the supercharging pressure is high enough to overcome the force of the valve springs. Otherwise, completely fill the fluid end of the mud pump through opened valve pots.

Maintenance of Triplex Pumps

Proper maintenance, of course, is the best way to keep the pumps operating efficiently, as well as to prevent unplanned downtime. Pump operating costs stay lower with a planned maintenance program than with a program that takes care of equipment only as emergencies arise. If the crew repairs the pumps when the rig is shut down for other activities (such as during a bit change or while waiting on cement), the cost of changing parts is minimal in terms of rig downtime. A cost-saving maintenance program means giving careful attention to how many hours the pump has been operating, inspecting it regularly, and replacing parts when they have been operating for a given length of time.

Similarly, using replaceable pump parts until they wear out completely is false economy. For example, the price of a piston is only one-fourth to one-sixth that of a liner. A derrickhand who leaves a leaking piston in place until it fails may have to replace the liner as well as the piston. Replace parts in groups that have similar life expectancies.

Triplex pumps are easier and less expensive to maintain than duplex pumps. For example, piston rods do not wear as quickly in a triplex pump because they do not have to move through a packing. Although the pistons and liners wear faster due to the high speed of operation, their lower cost and ease of replacement clearly outweigh the need to replace them more often. Lubrication and cooling are even more important to the life of triplex pumps because they operate at such high speeds.

Power End

Lubrication of the driving gears, shaft bearings, and crank throw bearings is essential to long pump life. Keep the lube pump, if the mud pump has one, running whenever the mud pump is operating at a slow speed. Keep an eye on the oil pressure and temperature, and check the oil level daily by opening check plugs or using a dipstick.

Keeping the lube oil clean is just as important as keeping the level correct. Most power-end troubles come from dirty oil because a mixture of oil and mud is very abrasive and can quickly ruin bearings, gear teeth, and crosshead slides. Inspect the oil-pressure gauge frequently to determine whether there are leaks in the system, and change the oil regularly. Clean the magnets whenever you change the oil, and check the pony rod wipers for leakage and wear. The pump may have a settling chamber (fig. 66). Remove the cleanout plates weekly and allow any water to drain off. Clean this chamber and the sump thoroughly when changing the oil.

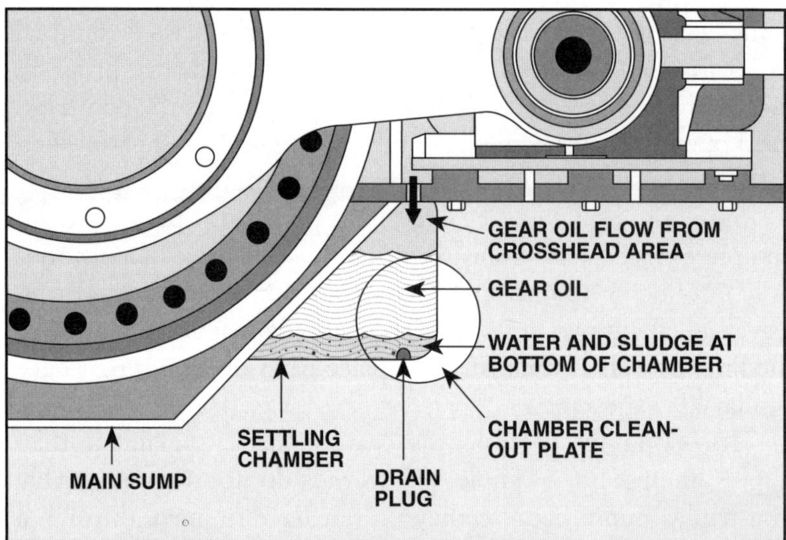

Figure 66. Settling chamber on pump's power end

Crosshead and Extension Rods. As the power end of the pump wears with normal use, the crew will need to adjust or replace some parts. A worn crosshead may cause uneven piston, rod, and liner wear. As described earlier, check the power end to find the cause of mechanical knocks. When bearings and gears are operated in a loose condition, they can develop knocks that may cause failure at a critical time on the job. If you notice excessive vibration or overheating, shut down the pump and get a mechanic or service person to investigate it.

Tighten the connections between the piston rods and pony rods to the recommended torque. If they are too tight, the connections may break, but the high-intensity load of the pistons' reciprocating movement can work them loose if the torque is too low.

Replace the extension rod wiper as needed to prevent mud from leaking into the power end of the pump. Inspect the extension rod crosshead locknut regularly for tightness.

Fluid End

Derrickhands should check the fluid end of the mud pumps regularly, at least every 15 minutes, while the pumps are working. They should—

1. check the pony rod clamps,
2. monitor the suction and discharge pressures,
3. listen for knocking in the valves,
4. look for fluid leaks, and
5. look for valve cap leaks and check tightness.

Derrickhands should also drain the piston leakage sump tank periodically, check the supercharging pump's gland packing and listen for strange sounds, check the suction dampener for proper pressure, and check the cooling pump and cooling system. They record all these checks in a log, as well as the number of hours on each piston and valve.

Liners

Liner life depends on piston clearance and operating pressure, the effects of corrosion, and the effects of erosion by sand and silt in the drilling mud. When corrosion is a problem, the drilling contractor can use corrosion-resistant liners to increase piston and liner life. A worn liner shortens the life of a new piston, so regularly check liners for wear by measuring the clearance, the depth of the wear groove on the piston, and for roundness.

In single-acting pumps, the back end of the liner is exposed and it is very easy to check the condition of the liner bores when the pump is stopped.

Clearance. Clearance is the distance between the liner and piston (fig. 67). It should be no more than 1/32 inch (0.8 millimetres) for high-pressure operation. A pump that operates at high pressure requires smaller clearances to hold back the pressure and prevent leaking. A piston and liner with a clearance of 0.04 inch (1 millimetre) will last no more than half as long as new parts with a clearance of 0.01 inch (0.25 millimetre). Table 9 is an example of the amount of wear that one manufacturer allows for various pressures.

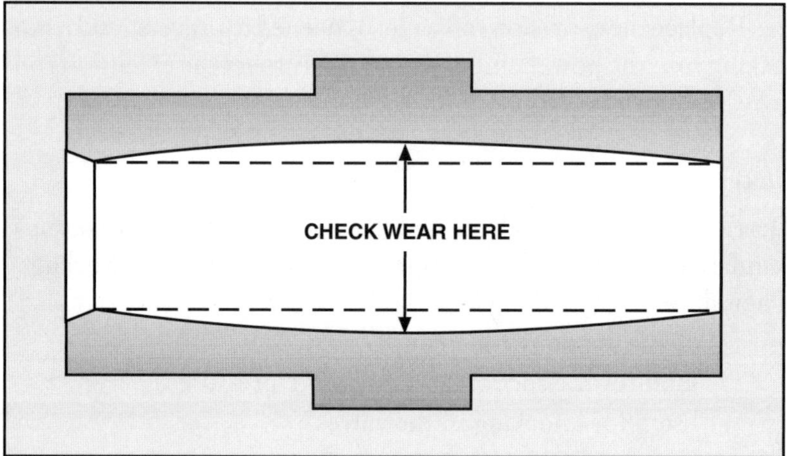

Figure 67. Measuring liner wear for a single-acting pump

The clearance can be too large, either because the piston's diameter is too small or the liner's diameter is too big. For example, the clearance may start out too large because someone did not install pistons and liners of the same size. Clearance may also increase when oil gets into the mud system, either from waste around the rig or from oil-bearing formations. Such oil contaminates mud and damages piston rubbers. The oil in oil-base or oil-emulsion muds may also damage rubber parts. To prevent such damage, derrickhands should be sure to install oil-tolerant synthetic rubber parts. Still another way in which clearance can increase is when abrasive sand and silt in the mud enlarges the diameter of the liner. To check for liner wear, check the diameter by measuring the bore.

Table 9
Allowable Liner Wear

Operating Pressure psi (kPa)	Liner Wear Limits inches (mm)
1000–2000 (6,895–13,790)	3⁄32–1⁄16 (2.4–1.6)
2000–3000 (13,790–20, 685)	1⁄16–3⁄64 (1.6–1.2)
3000–4000 (20,685–27,580)	3⁄64–1⁄32 (1.2–0.8)
4000–5000 (27,580–34,475)	1⁄32 max. (0.8 max.)

Clearance may also be uneven. Pressure in the bypasses or a fluid-cut area exerts a force that pushes the piston against the liner wall on the opposite side. This causes the flange of the piston body to wear away and appear to be flat. Any flatness or out-of-round condition increases the clearance between the piston and the liner bore. When the flange is worn to the bottom of the wear groove, replace it.

Cylinders

Check for fluid cutting, which occurs when leaks etch the metal. Clean the weepholes. If fluid is leaking from a weephole, the packing has failed. Examine the liner bore outside the stroke area for small pits or discoloration that are evidence of corrosion. Pitting may also be evident on other parts of the pump that the mud contacts, such as a valve or seat.

Inspect cylinder head studs regularly, and replace them if they are worn or broken. Replace cylinder head gaskets if they are not in good condition. Tighten cylinder head stud nuts evenly in a crisscross pattern to their rated makeup torque.

Pistons and Rods

A piston may fail all at once, or it may fail slowly. A piston that fails slowly allows fluid to slip by on each stroke. This high-velocity fluid cuts, or damages, the liner, causing its eventual failure. A liner costs four to six times as much as a piston, so derrickhands should make every effort to detect piston failure as soon as possible and change it immediately.

To check the pistons, disconnect the piston rods from the extension (pony) rods, and pull the piston and rod assembly from the liner. Usually the rubber is the part that needs to be replaced.

In triplex pumps, fluid pressure operates on one side of the piston only. Because pressure operates only on one side, manufacturers do not taper the bore of the rod where the piston body fits onto the rod. Because the piston and rod bores are straight rather than tapered, the pistons are easy to remove from the piston rods. A simple packing ring provides a fluid-pressure seal. *API Spec.* 7 gives details of piston and rod connections for single-acting pumps.

To install a piston onto the rod, take the following steps:

1. Make sure to remove all mud and scale from the piston body and rods.
2. Save the rod's shipping tube and put it over the crosshead end to protect it. If the shipping tube is not available, protect the crosshead threads by tying a rag securely over them.
3. Use a new, lightly greased packing ring between the piston body and the rod shoulder.
4. Slide the piston onto the rod, and oil the rod threads.
5. Before tightening the nut, make sure the piston fits squarely against the rod shoulder. Do not use the same nut more than three times. It loses its ability to grip the threads if reused too often.
6. Tighten the piston on the rod, but do not overtighten. Use no more than one person on a 5-foot (1.5-metre) wrench.
7. To replace the rod, clean the threads on the crosshead end and lubricate them before installing the assembly back in the pump.

The second and third pistons used in a worn liner will wear out faster at the same pressure than the first piston because the initial clearance is greater (fig. 68). Always replace the piston when replacing the liner.

Lubrication and Cooling

The piston rubbers in a triplex pump can burn and score the liner in less than 30 minutes if the piston and liner are not cooled. As pump speed increases or the mud temperature rises, proper cooling and lubrication become even more critical. Check that the cooling spray completely flushes the entire stroke area in the liner. This usually requires at least 5 to 10 gallons (20 to 40 litres) of coolant per minute flowing onto each liner. To determine the specific amount, refer to the manufacturer's recommendations. Test daily whether enough coolant is spraying and whether the nozzles are aimed correctly by carefully placing your hand on the liner—it should be cool to the touch.

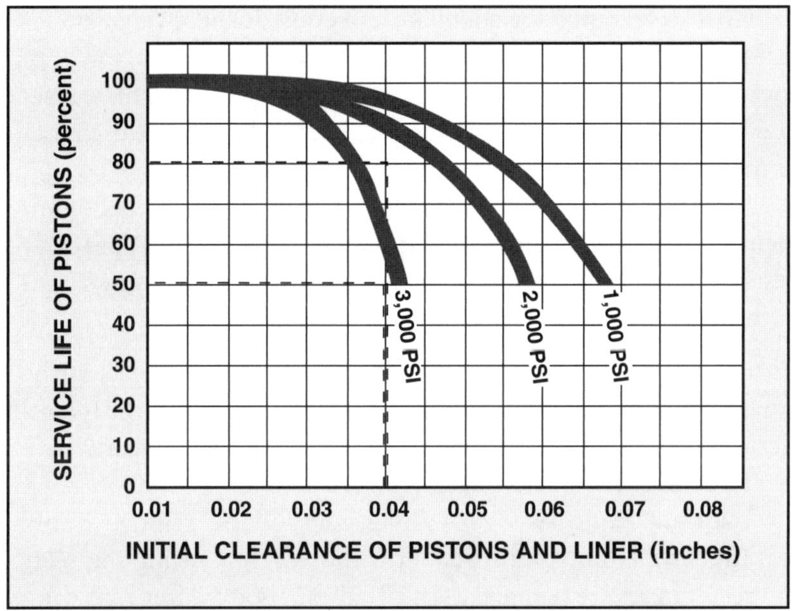

Figure 68. Piston service life related to piston liner demand

Liner Packings

Check the liner packings daily and tighten them, when necessary, to the manufacturer's specifications. Packings for liners with thin walls, ⅝ inch (16 millimetres) or less, can bottleneck the liner if tightened too much. Bottlenecking reduces the diameter of the liner where the packing is too tight. Bottlenecked liners can also result if derrickhands tighten liner packings when the pump is under pressure. Tighten liner packings only when the pump is not operating.

If the pump has a weephole, be sure that the hole in the liner packing lines up with the weephole in the pump body when the liner and packing assembly are in place and seated solidly.

After replacing liner packings, start the pump, allow it operate for several minutes, and check the packing for leaks and liner movement. Stop the pump and retighten the packing if necessary.

Valves

During times when the pumps are not operating, inspect the pump valves for fluid cuts and pits in the sealing area of the disk. Replace worn disks before the fluid washing damages the valve body or seat. After checking, always return the valve to the same seat it came from. Valves and seats tend to have matching wear patterns, so they will last longer if you keep them together.

Check clearance for adequate valve lift. Inspect valve guides and replace them if wear is evident. Worn guides allow valves to cock—that is, the guide moves inside the bushing. This causes uneven wear, possible sticking, and washouts. Check the diameter of the bushing with a micrometer.

Replace valve springs on schedule. If the manufacturer recommends a certain spring for a given valve, use that spring. A properly designed spring reduces fluid knocking and excessive pressure surges in the discharge. Weak springs cause fluid knocking.

Changing Seats and Cleaning Pots. Replace valves and seats at the same time when wear exceeds limits recommended by the manufacturer or when you notice fluid cutting. Do not use a worn valve in a new seat, nor a new valve in a worn seat.

Use a mechanical wedge or a hydraulic puller to remove the old seat (fig. 69). Do not cut seats out with a torch. Cutting out seats can result in severe damage to the fluid-end body. After removing the seat, clean the valve pot. Begin cleaning the pot at the sealing surface of the pot cover gasket and finish with the seat taper so that no dirt or grease falls onto the clean, dry taper. Then clean the pump deck of rust or dried mud.

Figure 69. Hydraulic valve seat puller

Check the seat by solidly pushing it down into the valve pot by hand. It should stick. If it does not, check the two tapers for complete contact between the taper of the seat and the taper of the valve pot. As soon as you have seated the seat by hand, drive it into the pot with a seat driver and a 16-pound (7-kilogram) sledgehammer. If no seat driver is available, place an old valve body in the seat and drive the seat in with a bar or drop something heavy on it. Listen to the sound this makes; the sound will change when the seat is properly seated. Do not rely on pump pressure to make this initial seal. If you do, corrosive drilling fluids can get into the taper and cut the surfaces of both the valve seat and valve pot.

Suction Equipment

If the pump has a suction strainer, remove it from the suction line when the pump is shut down and clean it. While the pump is down, clean the suction line and any suction-line valves and hoses. Also clean the manifold connections and be sure that they are free of solids that have settled out. Hard-packed sand or barite blocking part of the suction line can prevent flooded suction to the pump. Once derrickhands become familiar with a particular pump, they may be able to detect sediment buildup by feeling the suction line, which will be noticeably warmer where packed solids have not accumulated.

If the pump's suction line is reinforced rubber hose, inspect it regularly for collapse and for air leakage. Also, inspect all suction fittings, particularly unions, and replace and tighten gaskets if you suspect air leakage. When moving or rigging up, be careful to avoid kinking or cutting rubber suction hoses.

Maintenance of Duplex Pumps

A duplex pump has more parts to fail, and replacing them is more difficult, but just as necessary. This section discusses only parts that are not similar to those on a triplex pump. Maintain similar parts in a similar way.

Pistons and Piston Rods

In a duplex pump, failed pistons allow fluid to pass from one side of the liner to the other, which reduces volumetric efficiency. Piston rods in duplex pumps wear on the outside diameter as the rod strokes through the packing (fig. 70), so rig personnel need to replace them periodically.

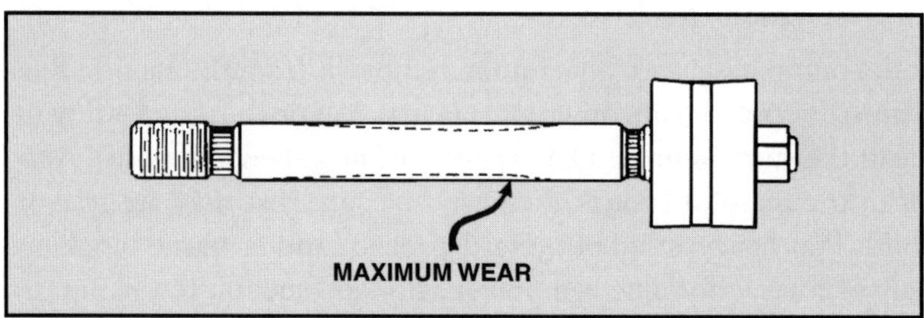

Figure 70. Typical rod wear pattern of a double-acting pump

Manufacturers taper the end of piston rods for duplex pumps. Similarly, they also taper the piston-body bore to fit the rod. The taper is necessary because fluid pressure occurs on both sides of the piston in a duplex pump. The taper, in addition to the threaded nut, ensures a very firm fit. To connect the rod taper to the piston taper, push the piston onto the rod. It should move solidly and stick firmly onto the rod taper. The flange of the piston widens about 0.002 to 0.004 inch (0.05 to 0.1 millimetre) as you force the piston onto the taper. To finish installing the piston on the rod, use the same procedure as with a triplex. That is, check that the piston is properly seated, carefully thread the nut on the threaded portion of the rod, and tighten it to the manufacturer's recommended torque.

Rod Packing and Stuffing Box

The stuffing box and packing wear more quickly if the rod packing is too tight and if they overheat. For this reason lubrication and cooling are very important to the life of rods and packings. Never leave worn-out rod packing in a pump. Always replace the rod packings at the same time as the rods.

Stuffing Box. Keep bushings and spacers in the stuffing box in good condition. Once the box is worn, especially in the brass bushings or sleeves, new sets of rod packing deteriorate rapidly. Replace the box when necessary.

Check that the piston rods are centered in the stuffing boxes. If they are not, they will deform the rod packing and reduce the life of the rod and packing. Check the piston rods regularly for overheating and leaks at the point where they enter the stuffing box.

Installing Rod Packings. When installing or tightening the rod packing, follow the manufacturer's recommendations closely. If the nut is too loose, it may leak, which can wash out the rod, rod packing, or stuffing box. If the nut is too tight, it compresses the packing, which binds and scores the rod. Overtight packing nuts are the usual cause of trouble in the piston rod packing. Tighten the nut so that the pressure on the packing is moderate on a smooth rod. Two types of packings, lip and chevron rod packings, use pump pressure to aid in sealing, so require much less packing pressure than solid packings.

Most packings have a split to make installation and removal around the rod easier (see fig. 58). If the packing set contains more than one split ring, line up the rings so that no two splits come together.

Lubrication and Cooling

The lubricating and cooling system for rods may be either a fitting that splashes or drips oil onto the rod, or a pressure lubrication system that uses a small lube pump.

As in triplex pump cooling and lubricating systems, keep an eye on the oil level and pressure and the aim of adjustable nozzles. If you have to add lubricant too often, look for a loose fitting or a leaking hose. If a leak occurs, replace the packing and examine the rod for grooves. Clean the settling pan regularly to prevent abrasives from clogging the pump inlet and recirculating through the lubricating system, which wears the rod and packing.

Because the rod is open to the air, the lubricant also cools the rod and flushes it to clean off any dust or dirt that may have accumulated.

To summarize—

- Maintain good suction and discharge flows by using a supercharging pump and keeping piping as short and straight as possible.
- Keep the pressure-relief valve and suction and discharge strainers clean.
- Always prime the pump before starting it up.
- Continuous maintenance—monitor the pony rod clamps, monitor the suction and discharge pressures, listen for knocking in the valves, look for fluid leaks, and look for valve cap leaks.
- Daily maintenance—check the oil and coolant levels; inspect the fluid-end body, liners, pistons, packings, valves, and suction lines and connections; and check the dampeners.
- Regular periodic maintenance—replace worn pistons immediately when they begin to fail, replace other parts on a schedule.

Centrifugal Pumps

In addition to the two or more large mud pumps, a rig usually has several smaller *centrifugal pumps* for auxiliary purposes (fig. 71). A centrifugal pump has no pistons; it transfers energy to a liquid through action of a *rotating impeller* (fig. 72). The impeller turns on a drive shaft inside a housing, or casing, which may have a replaceable liner.

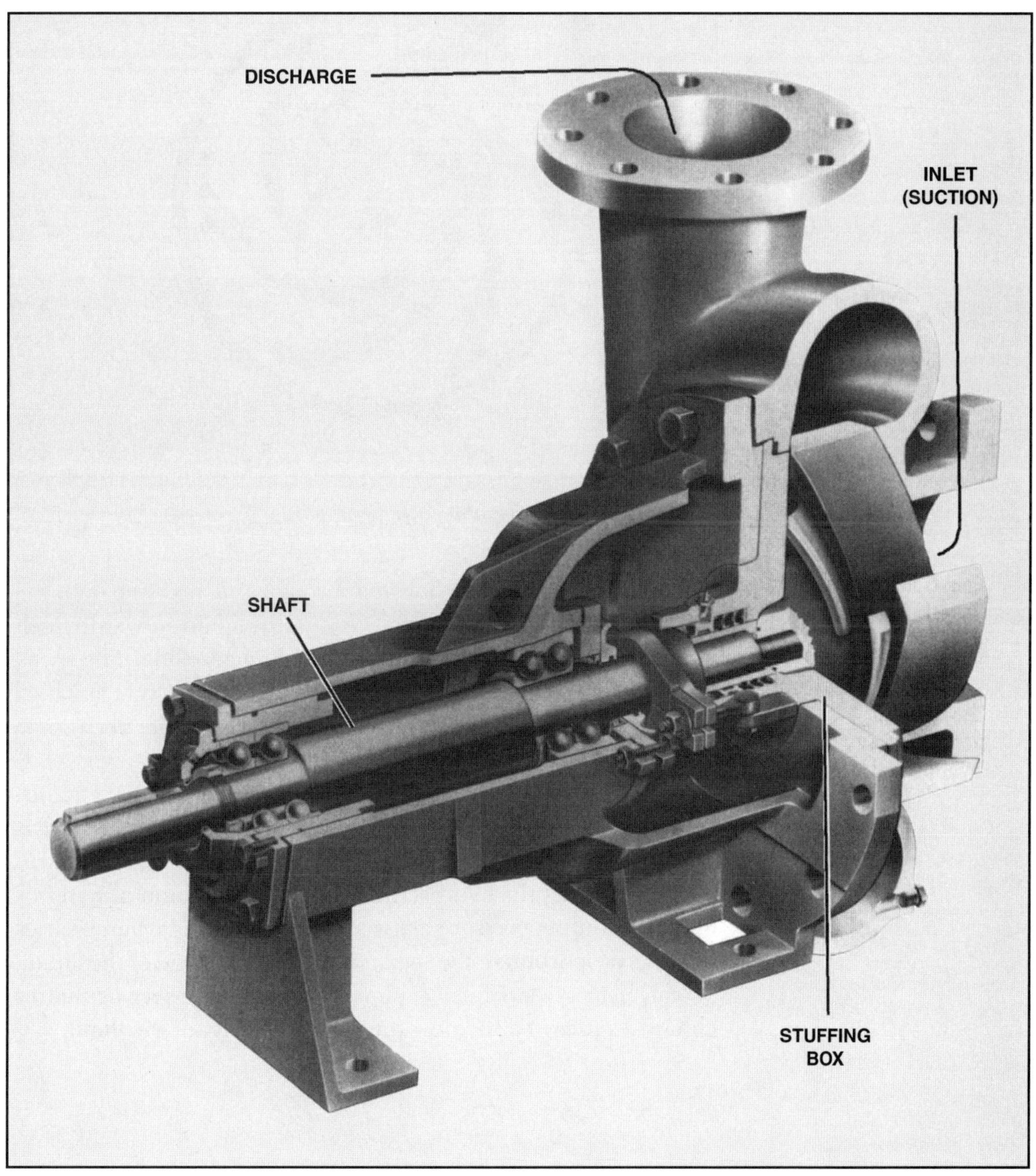

Figure 71. Cutaway of a centrifugal pump (Courtesy of National Oilwell, Mission-Fluid King)

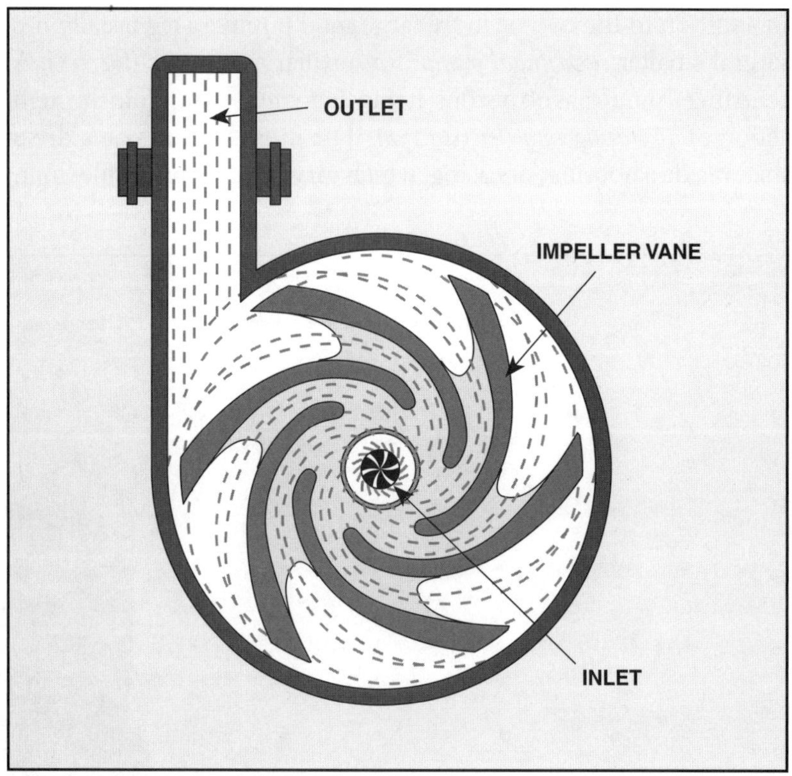

Figure 72. Operation of a centrifugal pump

The liquid flows into the housing through a suction line onto the spinning impeller's vanes. The vanes hurl the liquid outward, which increases its speed and forces it out the discharge line. The speed is converted to a pressure much higher than the pressure at which it entered the pump. An electric motor or drive belts from some other prime mover drives the shaft of the pump.

Rig personnel often express the pressure output of a centrifugal pump in feet or metres of head rather than psi or kPa. *Head* is another word for pressure, and the calculation of feet or metres of head includes the hydrostatic pressure of a column of water or mud. The output pressure in psi (kPa) of a centrifugal pump varies in direct proportion to the specific gravity, or weight, of the liquid. So calculating centrifugal pump requirements in feet or metres eliminates the need to know the specific gravity of the fluid.

Advantages of Centrifugal Pumps

Rig personnel use centrifugal pumps for many purposes: supercharging the mud pump; moving water for making up mud, mixing cement, washing down equipment; cooling the brakes; feeding mud to conditioning equipment; and so on. Centrifugal pumps are better suited to these jobs than reciprocating pumps because they are smaller, quieter, and less expensive. Also, they can move large volumes of liquid at low pressures. Conversely, reciprocating pumps are better for moving small volumes at high pressures.

Centrifugal pumps do not produce the pulses and vibrations in piping that reciprocating pumps produce. What is more, solids that remain in the mud after it passes through the shale shaker do not affect centrifugal pump performance.

Fluid Regulation

Derrickhands can regulate the volume of fluid that a centrifugal pump delivers by (1) changing the speed of rotation, or (2) throttling (opening up or closing down) the discharge valve. Changing the speed is the better method to use, because throttling the discharge valve always wastes power, and, ultimately, the fluid erodes the throttling valve.

Crew members can totally stop the flow of fluid from a centrifugal pump by closing the discharge valve. Manufacturers do not, however, recommend this practice because it can shorten the life of the impeller and housing. Running the pump with the discharge line shut off causes the fluid in the pump to overheat. If you cannot shut down the pump to stop pumping fluid, rather than closing the pump's discharge valve, rig up a special discharge line. Run the special line to the suction tank from which the pump draws fluid. With this system, the pump recirculates fluid back to the suction tank, which allows the pump to idle.

Operating Centrifugal Pumps Efficiently

Preventing air from entering the pump is as important in centrifugal pumps as it is in reciprocating pumps. Air trapped in the suction line can cause the pump to cavitate—that is, to fail to get enough fluid in the casing to fill it up. Air drawn in through the suction piping can also cause the pump to lose prime on start-up.

One way to ensure flooded suction is to deeply submerge the suction line in the suction tank. The fluid level in the tank must stay well above the end of the suction line. If the pump must operate with a low fluid level in the tank, use an oversized suction line. Also, suction piping should slope upward from the liquid source to the pump to avoid traps that will accumulate air. Do not arrange the suction and discharge piping so that fluid enters the tank above the pump suction. The splashing of the mud falling from the pipe creates air bubbles, which then enter the suction line (fig. 73). To do it right, submerge the discharge piping in the tank. Avoid any practice that mixes air or gas into the drilling mud (like running mud agitators close to the centrifugal pump).

Because a centrifugal pump produces almost constant pressure, the horsepower requirement depends on the volume. If the motor driving a centrifugal pump overloads, act immediately to lower the volume by partially closing the discharge valve to decrease the flow, which lowers the horsepower required.

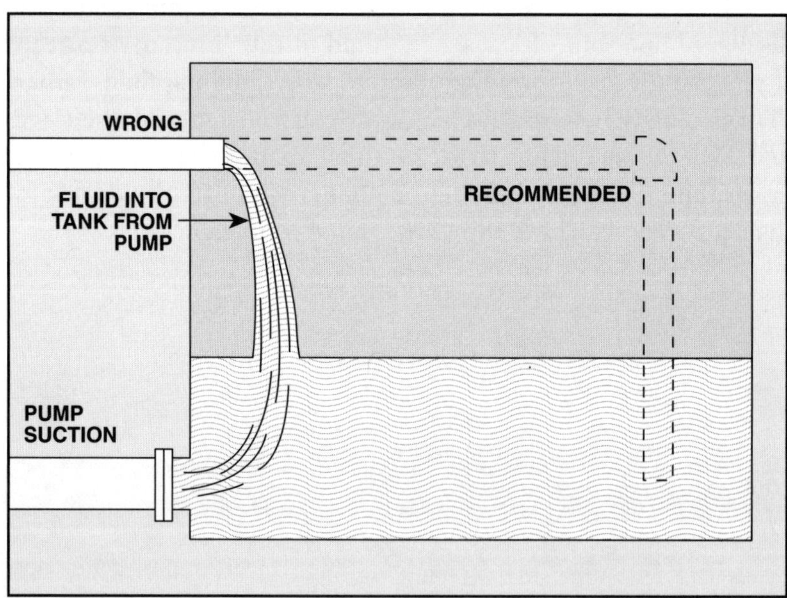

Figure 73. Suction and discharge piping for a centrifugal pump

Pump Sizing and Selection

Manufacturers rate centrifugal pumps by size of the impeller: a pump with a larger impeller can move a larger volume of liquid. The contractor selects the proper size of pump for each job. Supercharging pumps, for example, must produce enough pressure to fill the reciprocating pump liners, but not so much that the suction valves remain open too long, which reduces the life of the valves and valve seats. A pressure-reducing valve allows the derrickhand to regulate pressure. Table 10 shows typical impeller sizing for various applications.

Table 10
Impeller Sizing

	Speed of 1,150 rpm		Speed of 1,750 rpm	
Application	Maximum Size inches (millimetres)	Minimum Size inches (millimetres)	Maximum Size inches (millimetres)	Minimum Size inches (millimetres)
Desander	14 (355)	13 (330)	10½ (267)	9 (229)
Desilter	14 (355)	13 (330)	10½ (267)	9 (229)
Mud mixing	14⅛ (359)	13 (330)	10½ (267)	9 (229)
Degassing	14⅛ (359)	12 (305)	10½ (267)	9 (229)
Supercharging	13 (330)	11 (279)	N/R (N/R)	N/R (N/R)
Brake cooling	N/R (N/R)	N/R (N/R)	13 (330)	10½ (267)

Maintenance of Centrifugal Pumps

Any part of a centrifugal pump may need to be replaced at some time during its life. In normal operation, wear varies directly with speed. Figure 74 shows a troubleshooting guide and a checklist for starting up a centrifugal pump.

Lubrication and Bearings

Centrifugal pumps use roller bearings between the drive shaft and the impeller, which may be lubricated by either oil or grease. Do not overlubricate them. Too much oil makes the bearings run hot. Too much oil or grease retains the heat bearings generate as the unit operates. Overlubricating bearings can be as bad as underlubricating them. On pumps that have a grease seal between the shaft and bearings, adding too much grease can allow some to fall out, leaving a visible opening in the seal. Grease on bearings does not become contaminated as easily as oil, so regreasing the pump should not be necessary very often. When contamination does occur, disassemble the pump and remove the old grease. Then pack the bearing by hand and fill the bearing cap approximately one-third full with clean grease. When using oil, keep it clean and change it on a regular schedule.

With either type of lubrication, periodically check for overheating by cautiously feeling the outside of the bearing housing. Such checks may prevent serious damage. Pump bearings can operate at temperatures up to 160°F (71°C). Any temperature above 160°F (71°C) is too hot for the hand for longer than 5 seconds. An increase in noise or a sudden temperature rise (above 200°F, or 93°C) may indicate that a bearing is beginning to fail. Replacing the bearing before it completely fails prevents damage to the impeller, shaft, and housing.

Stuffing Box

To repack the stuffing box, which you can do without disassembling the pump, remove all the old packing and clean the box. Place three rings of packing in the bottom of the box so that, when compressed, the lantern ring is in its proper location beneath the sealing tap. Stagger the ring joints. Draw up the packing rings snugly by tightening the gland. Pack the remainder of the box, draw up the packing snugly, and back off the gland until it is finger-tight. Packing expands with heat, and a box that is more than finger-tight when cold usually smokes on start-up. The stuffing box should allow a small amount of liquid to drip. The dripping liquid cools the packing, so do not tighten the gland to the point where it prevents any liquid from dripping. Tighten the gland a half turn at a time until the packing allows liquid to drip at the desired rate.

CAUSES	SYMPTOMS						
	Noise vibration	Not pumping	Not delivering enough liquid	Not enough pressure	Uses too much power	Starts, then loses suction	Bearing wear
Pump not primed		●	●				
Speed too low		●	●	●			
Discharge head too high		●	●				
Suction lift higher than design	●	●	●			●	
Impeller completely plugged		●	●			●	
Wrong direction of rotation		●	●	●			
Plugged suction or discharge line		●	●				
Insufficient suction head for hot liquid			●				
Foot valve or suction line not immersed deeply enough			●			●	
Impeller damaged			●	●			
Casing packing defective			●	●			
Impeller diameter too small				●			
Excessive amount of air or gas in liquid				●		●	
Speed too high					●		
Total head lower than design					●		
Specific gravity or viscosity too high					●		●
Bent shaft	●				●		●
Faulty electric motor wiring and voltage			●		●		
Rotating elements bind	●				●		●
Leaky suction line or shaft seal		●	●			●	
Misalignment	●						●
Bearings worn	●				●		●
Rotor out of balance							●
Excessive internal thrust							●
Lack of lubrication, dirt, or excessive cooling	●						●
Suction or discharge piping not anchored	●						
Improper foundation	●						

CHECKLIST FOR START-UP

1. Coupling aligned
2. Pump full of fluid
3. Suction valve open
4. Water on stuffing box (in case of double seal)
5. Oiler full (if oil lubricated)
6. Pump rotates freely by hand

Figure 74. Troubleshooting guide for centrifugal pumps

Check the adjustment of the packing gland daily. Too loose, and the packing leaks too much. Too tight, and it scorches, which can damage the shaft.

Alignment

The shafts that connect the motor and impeller must line up in all directions. Misalignment damages pump and motor bearings, packing, pump-to-motor couplings, and mechanical seals. Misalignment also causes the pump to lose power. Maintain alignment by carefully measuring the distances of each shaft up, down, and side to side from a stationary point. Then adjust as necessary.

To summarize—

- Rigs usually have several centrifugal pumps for auxiliary pumping needs.
- Centrifugal pumps are smaller, quieter, and cheaper than reciprocating pumps, and move large volumes of mud.
- Do not run the pump with the discharge line shut off; this causes the pump to overheat.
- Lubricate the bearings regularly, and check for overheating.
- Check the packing gland for tightness daily.
- Be careful to maintain alignment of the connecting shafts.

Other Circulating Equipment

Drilling fluid returning from the wellbore contains drilled cuttings, sand, other particles from the hole, and solids added purposely. Almost all circulation systems use a shale shaker to remove cuttings from the mud, but removal of finer solids is up to the operator. Mud with too many solids—drilled solids plus additives—becomes too heavy, or dense, which slows the drilling rate. Drilled solids can also increase the mud's thickness (viscosity). It takes high pump pressures to move thick, viscous mud. Also, muds that are too viscous may create other downhole problems that interfere with efficient drilling.

To control the viscosity at the level that is best for the drilling conditions, the operator may thin the mud. The three options for thinning mud are to use a chemical thinner, to add water, or to use equipment to remove the unwanted solids. The decision is based on cost. In a weighted mud, which may cost ten times more than an unweighted mud, operators usually employ mechanical equipment to remove solids. Mechanical equipment is usually the least expensive option.

Removing the most solids possible makes it easier to keep the mud in good condition. The cost for mud chemicals is less, and the overall cost of operation is lower when fine solids are not left in the mud. To this end, after passing through the shale shaker, drilling mud may pass through a desander, desilter, mud cleaner, and centrifuge to clean out smaller particles. The drilling contractor may also need to use a degasser to remove entrained gas.

The derrickhand usually maintains and adjusts the solids control equipment to keep the mud flowing properly.

Shale Shakers

Sometimes, if the viscosity of the drilling mud is low, the mud flows into a large settling tank where most of the solids can settle out by gravity. Usually, however, the mud passes over one or more shale shakers that separate out the cuttings.

How It Works

A shale shaker (fig. 75) consists of one or more vibrating or rotating screens. The screens have openings that are large enough to let the mud and its weighting material fall through, but small enough to hold back the cuttings. A small electric motor powers belts or gears that move the screen.

Tension bolts stretch the screen in a shale shaker tightly inside a metal frame. The derrickhand can adjust the angle of the screen both downhill and uphill to control how fast the mud flows over it. The rotating or vibrating mechanism moves on bearings and gets power from its own motor through drive belts.

Crew members should not allow contaminated mud to pass through the *shaker screen*. Cement and salt water are two common mud contaminants that can quickly plug a shaker's screen. Rig owners therefore often install a swing line or a *bypass gate* in the mud return line (the line in which mud flows from the well to the mud tanks).

Figure 75. Shale shaker (Courtesy of Harrisburg)

By adjusting the gate derrickhands can divert contaminated flow through a bypass line. The bypass line goes into the reserve pit or into a container for proper disposal later. When the mud runs clean again, they direct the flow back through the main return line to the shaker.

Mud Flow

A receiving tank at the end of the mud return line, the *possum belly*, helps control mud flow over the shale shaker. The mud flows into the bottom of the possum belly and travels over baffles before reaching the shaker. On the possum belly are three or four *flow gates* for each shaker. A crew member can lift them up to prevent flow to one shaker or another or to regulate the mud flow. When the bypass gate between two shakers is open, it dumps the mud in the possum belly so that it totally bypasses the shakers. The properties of the mud and the design of the equipment affect a shale shaker's screening capacity—that is, how fast the mud can flow through it.

Equipment

The smaller the mesh size of the screen, the slower the mud passes through the screen and the faster it plugs up. Manufacturers have come up with various improvements on the basic design of shakers and screens that allow a faster flow rate while still removing the largest amount of cuttings. One is a double-deck or triple-deck shaker. In this type, mud passes through an upper screen with a coarse mesh and then through one or two lower screens with a finer mesh. The upper deck removes the larger cuttings that could plug up the finer-mesh lower screens. New screen designs have also helped. Many shale shakers feature linear motion, which is a great improvement over older style elliptical-motion shakers. Linear-motion shakers do not tumble the cuttings as much as elliptical-motion shakers. Tumbling can make small cuttings out of large ones. Small cuttings may pass through the screens and enter the mud as undesirable solids.

Mud Properties

The mud's viscosity and the percentage of solids affect screening capacity. The more viscous the mud is, the slower it flows and the more time it takes to screen out the cuttings. Oil-base and invert-emulsion muds generally have higher viscosities than water-base muds of the same weight, so fine screens cannot handle as much oil mud as they can water-base mud.

Shakers also cannot handle as much mud to which crew members have added synthetic polymers to build viscosity. Shear stresses do not thin polymer muds. *Shear stress* is a force on the mud caused by its flowing movement. Shear stress forces gelled mud to flow. *Shear thinning* is a phenomenon in which the mud's viscosity decreases because of shear stress. In other words, when shear stress causes the mud to flow, the mud thins. Mud engineers call it shear thinning. Some polymer additives develop viscosity in the mud, but, when pumped, the mud does not thin. Put another way, polymer muds do not get thin, even when pumped fast.

Screens

Standard flat shaker screens are made of heavy wire, and most commonly have relatively large square, rectangular, or hexagonal openings, from 12 to 20 mesh. Table 11 shows the particle sizes that different mesh sizes screen. For example, the coarsest, a 12-mesh screen, sifts out 0.0601-inch, or a little less than 1⁄16-inch (1,540-micron, or 1.54-millimetre), particles. A 20-mesh screen removes 0.0298-inch (765-micron, or 0.765-millimetre) particles.

Table 11
Screens for Various Particle Sizes

Screen Mesh	Particle Diameter (microns)	Particle Diameter (inches)
12	1,540	0.0601
14	1,230	0.0483
16	1,020	0.0403
18	920	0.0359
20	765	0.0298
40	510	0.0199
60	250	0.0098
80	177	0.0069
100	149	0.0058

Drilled solids that have broken or disintegrated into fine particles pass through the standard-mesh screens. Fine-mesh screens on the shale shaker or cleaning equipment farther downstream can remove these finer solids. A rule of thumb for determining the right size mesh is that the mud should travel down one-half to two-thirds of the length of the screen before completely passing through it.

Fine-Mesh Screens

Fine-mesh screens, 60- to 250-mesh, screen out much smaller particles, but their use presents special problems. The smaller the mesh size, the slower the mud passes through the screen and the faster it plugs up. Standard shaker screens have an open area of 50 to 55 percent, whereas an 80-mesh screen has an open area of only about 30 percent. To handle the same volume of fluid, the finer mesh screen has to be about 50 percent larger in total surface area.

Another disadvantage of fine-mesh screens is that the wire itself is smaller and not as rugged as larger-mesh screens. To alleviate this problem, the shaker may use a steel reinforcement grid between the supports and the fine screen to keep the metal supports from cutting the screen.

An improvement is the sandwich design. These fine-mesh screens consist of two screens, one above the other, supported by a backing screen and a plate with large perforations. These screens do not plug up as easily as older designs, and last longer because the backing screen supports the two fine-mesh screens.

The latest screen is a corrugated sandwich design called a pyramid screen. It has 40 percent more surface area than a flat screen, which almost doubles the screen's flow capacity. The cuttings tend to fall into the troughs, keeping the raised portions open for the drilling mud to pass through (fig. 76). Because it can handle a faster flow rate than traditional screens, the mesh can be finer. Pyramid screens are particularly valuable when drilling with oil muds because the cuttings falling into the waste pit are dryer. Dryer cuttings mean that less of the expensive oil mud goes to waste, and the waste contains less environmentally hazardous oil to dispose of.

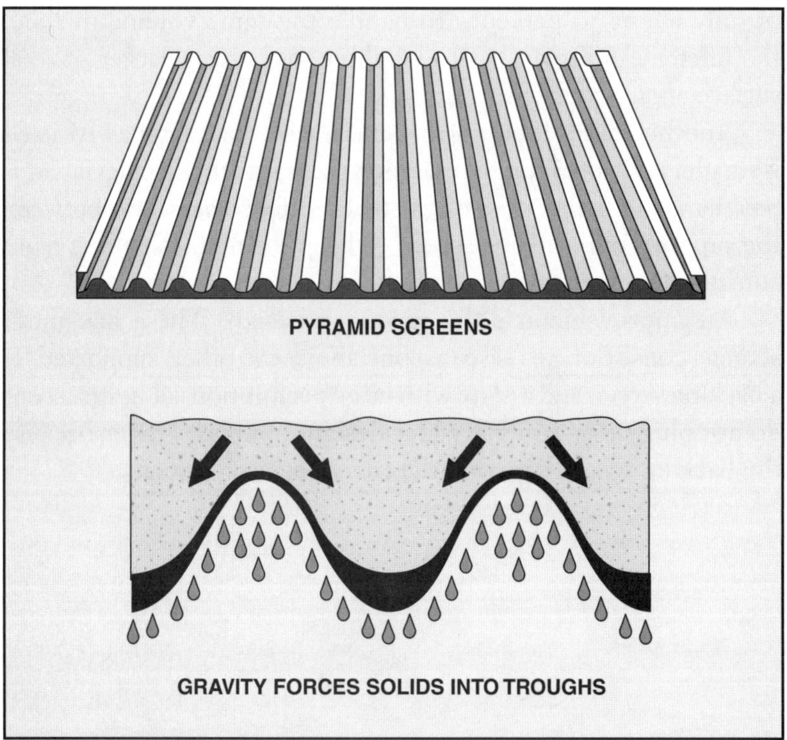

Figure 76. Pyramid screen

Sand Trap

Usually, crew members install normal screen sizes when drilling the top part of the hole. Normally, hole size is large and drilling rate is fast. Normal-size screens allow sand and finer solids to pass through. Some sand easily settles out of the mud. So, the tank that receives the mud first after it leaves the shaker is called the sand trap because of this settling action. The bottom of the sand trap slopes downward toward cleanout openings. The solids accumulate here, and the crew dumps them out periodically.

The sand trap is an important part of the separation system. If the screen becomes damaged, the sand trap catches the larger particles that may plug or damage downstream mud conditioning equipment. If the shaker bypass is accidentally left open a few minutes, the sand trap can catch and hold the large particles for a short time. It also acts as a crude desander below the vibrating screen when the mud is heavily loaded with sand.

Plugging

The main way a shale shaker can malfunction is that the screen can plug up with solids. Dissolved salts, grease, and fine solids in mud are possible culprits.

Salts dissolved in water-base drilling mud such as common table salt (sodium chloride), *anhydrite* (anhydrous calcium sulfate), gypsum (another form of calcium sulfate), and carbonates can cause serious screen plugging. If the rig crew does not wash down the screen at regular intervals, especially before a round trip, the dissolved salts coat the wires as the water evaporates.

Once the salts harden on the screen, they are more of a problem to remove. The crew can remove water-soluble salts by washing the screen thoroughly with water, but calcium salts do not dissolve easily in water, and must be removed with acid.

Pipe dope or tool joint compound and greasy sediments found in heavier crudes can cause plugging when the temperature of the circulating mud is low. Generally, these greases do not present problems when the mud temperature is above 150°F (66°C). The crew can remove grease by washing down the screen with kerosene, diesel oil, or other suitable solvent. Be sure to provide catchment trays or other suitable containers to prevent the solvent from contaminating the area. Be sure, also, to wear and use proper protective clothing and equipment.

When using a fine-mesh screen, plugging is inevitable. Sooner or later, the grain size of the drilled formation will be small or large enough to create problems. Because no one can do anything about the size of sand grains or cuttings, the quickest solution is to change the screen mesh by using either the next smaller size (preferably) or the next larger size.

Maintenance

Maintaining shale shakers is relatively simple. With older shakers, crew members should regularly hose them down and use a wire brush to clean the screen. Crew members should not, however, hose down or wire brush newer units. Also, never spray any type of screen with water when using oil-base or invert-emulsion muds. The same warning holds true for muds containing large amounts of barite. In both cases, water ruins the mud.

Grease the shaker's bearings daily. Check the tension bolts that hold the screen in place regularly for tightness. Keep the screens pushed all the way to the back of the screen boxes to keep solids from leaking. When using only one screen on a double-deck shaker, put it on the bottom. Check the condition and tension of the drive belt weekly. Check the bypass gate and flow directors occasionally to make sure that they move freely.

For most efficient removal of solids, do not spray the shaker screen with water during circulation. Many fine particles adhere to the coarser ones. Spraying allows them to pass through the screen and remain in the mud rather than staying attached to the coarser solids that go to the reserve pit.

To summarize—

- The viscosity of the mud, its solids content, and the design of the equipment affect a shale shaker's screening capacity.
- Adjust the mesh size and screen angle to match the solids size and viscosity of the mud.
- When using water-base mud, wash down screens regularly to remove salts and grease that can plug them.
- Grease the bearings daily; check the drive belt, tension bolts, and bypass gate regularly.

Solids Control

Operators strive to control the amount, size, and type of fine *solids* in a drilling mud, especially weighted muds. Low solids mud generally gives faster penetration rates, creates fewer downhole problems, and reduces mud expenses. Solids control can therefore be vital to a good mud program.

When barite accounts for a mud's high solids content, crew members must reduce the amount of clay solids (silt) to keep the mud from becoming too viscous. But the size of the desirable barite is the same as the size of undesirable silt, and screens separate by size, not type, of particle. Because screens cannot distinguish particle size, operators have to use special conditioning, or solids removal, equipment. Conditioning equipment separates particles by their specific gravity. Barite is a high-gravity solid, and the drilled solids and bentonite are low-gravity solids.

In addition to the shale shaker, mud solids removal equipment includes desanders, desilters, mud cleaners, and centrifuges. Figure 77 names the various solids in mud (such as barite and sand) and their size, and indicates the conditioning equipment (such as desanders and mud cleaners) that operators can install to remove them. Generally, mud flows from one piece of equipment to the next (right to left in fig. 77), each of which removes smaller particles than the one before it.

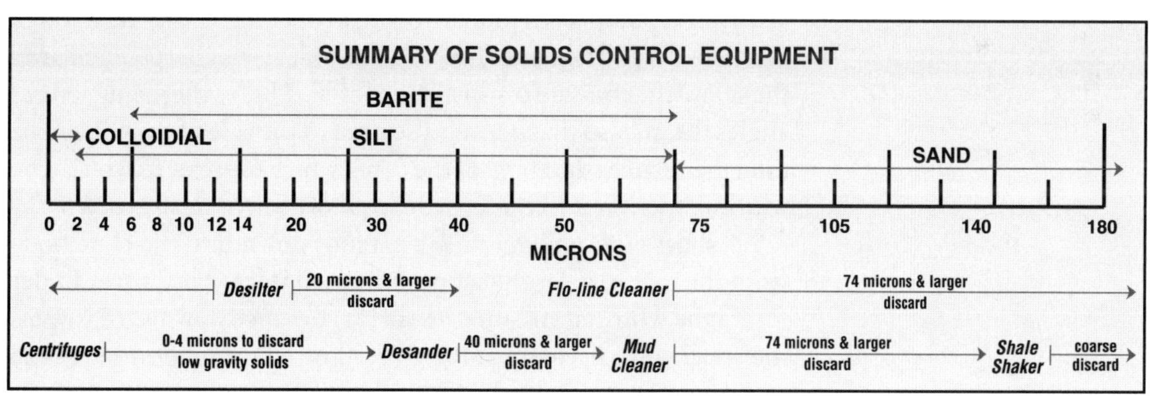

Figure 77. Solids size and removal equipment

Desanders and Desilters

Desanders and *desilters* remove solids from the mud that are small enough to pass through the shale shaker. As you might expect, desanders remove sand, and desilters remove finer particles, or silt. More specifically, desanders remove solids that range in size from 40 to less than about 74 microns. (Generally, shale shakers remove particles from 74 microns and up.) Desilters remove solids that range in size from about 20 to 40 microns. Mud cleaners remove solids down to about 7 microns in size, and centrifuges remove particles down to about 4 microns in size. (Remember, a micron is a small unit of measure—one millionth of a metre or 0.000039 inches. Mud solids are very small.)

Desanders and desilters use *hydrocyclones*, or *cones*, to separate out the sand or silt. Desanders and desilters extend the life of pump liners, pistons, and rods because they remove particles that easily erode such equipment.

How a Hydrocyclone Works

A hydrocyclone is a durable plastic or ceramic cone (fig. 78) with an inlet on one side, an outlet for liquid at the top, and an outlet for solids at the bottom. A hydrocyclone is simple, inexpensive to operate, and has a relatively high capacity (it can desand, desilt, or clean a lot of mud). It has no internal moving parts. In operation, a centrifugal pump feeds mud through an opening in the large end of the cone-shaped housing (fig. 79). This inlet is not perpendicular to the side of the cone; instead, it is tangential. That is, the manufacturer angles the inlet so that it sends the mud to the side of the cone. The mud then strikes the curve of the cone, which causes it to whirl. The whirling motion is like a tornado or, more scientifically, a *vortex*.

A short pipe, a *vortex finder*, extends down into the cone body from the top, to a depth just past the feed inlet. The vortex finder forces the whirling stream (the vortex) to move downward toward the small end (the apex) of the cone. The whirling motion creates centrifugal force, which throws the larger, heavier particles outward toward the wall of the cone. At the same time, the finer, lighter particles, which move outward more slowly, do not separate but remain as part of the fluid mud.

The larger particles and a small amount of fluid continue moving to the bottom of the cone and pass out through the apex. This lower discharge stream is the *underflow*. The remainder of the fluid, containing the smaller particles, reverses direction and spirals back up inside the whirling mud stream, leaving by way of the vortex finder. This stream is called the *overflow*, or *effluent*.

Figure 78. Plastic hydrocy-clones (Courtesy of Sundowner Offshore Services, Inc.)

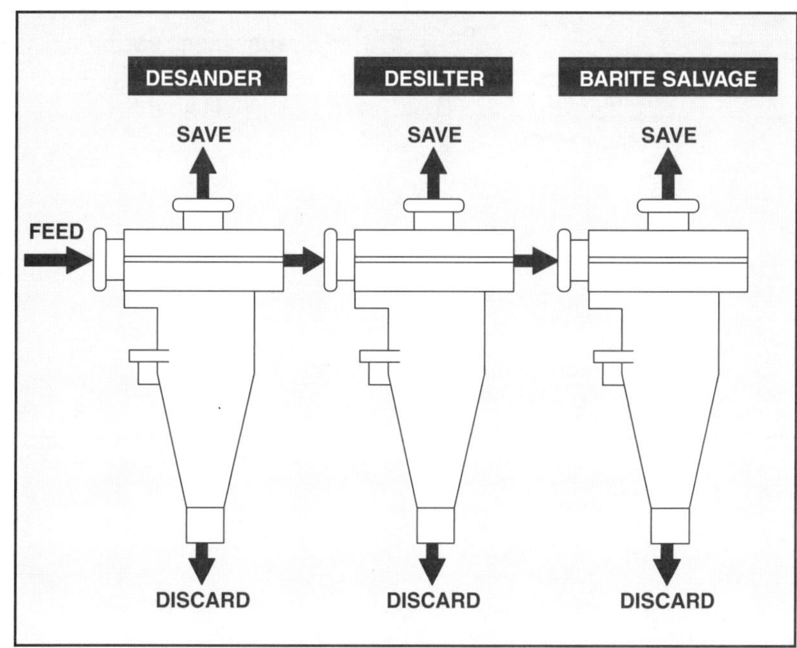

Figure 79. Operating principles of a cone-shaped centrifuge

Operators can use hydrocyclones as a desander, a desilter, or a barite scavenger (fig. 80). The main difference between the three is the size and number of the cones in the unit. The smaller the cones and the more of them in the unit, the finer are the solids they remove.

When operators use hydrocyclones as a desander or a desilter, the underflow contains the coarse solids to be discarded, while the overflow returns to the active mud stream. When they use a hydrocyclone to salvage (recover) barite, the underflow contains the barite weighting material and so returns to the active mud system, and the overflow contains the waste particles.

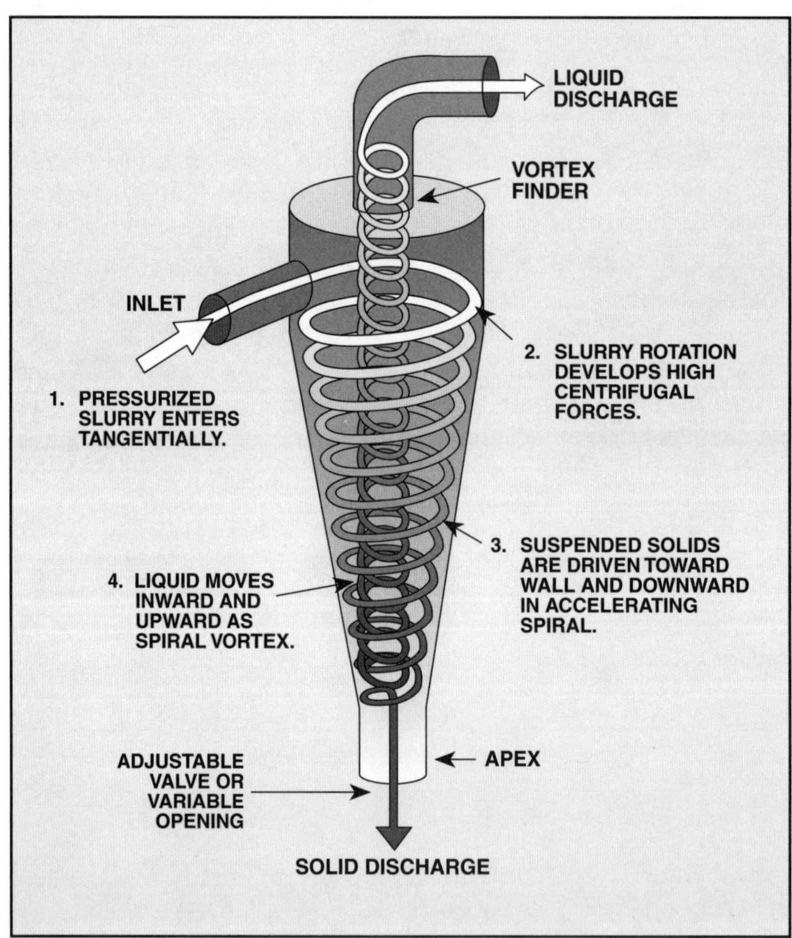

Figure 80. Applications of a cone-shaped centrifugal separator

Desanders

Desanders can be very useful early in the drilling of a well. At this time a lot of relatively large solids build up in the mud because of the fast drilling rates and a short settling time. Crew members usually install the desander downstream from the shale shaker. The desander discharges clean mud into the next tank. The desander's underflow, which holds the sand, enters a line that goes to the reserve pit. Desanders usually have one to three cones, which the manufacturer can mount at just about any angle on the unit's skid, from upright to 20 degrees above horizontal (see fig. 78). The vortex in a cone usually generates enough centrifugal force so that gravity does not have too much effect on the cone's ability to remove solids.

The size of the cone determines how much mud it can handle and what size solids are separated. A large cone allows a faster rate of mud flow, but separates out only the coarsest particles. Desanding cones are thus larger than desilting cones. Desanding cones have an inside diameter of 6 to 12 inches (152 to 300 millimetres), compared to desilting cones, which range from 3 to 5 inches (76 to 127 millimetres) in diameter.

Desilters

Good desilters, properly operated, reject all material of sand size, a high percentage of solids larger than 10 to 20 microns, and decreasing percentages of materials down to 2 or 3 microns. Many rig personnel erroneously believe that sand particles are the only abrasives in a mud system. Nothing could be further from the truth. Finer solids can be very abrasive. The angularity and hardness of a particle, not its size, determine how abrasive it is. Desilting the drilling mud can drastically reduce mud pump wear, hole problems, bit wear, time required to drill the hole, and the amount of water and chemicals required for mud treatment. Most operators recommend installing desilters in any mud system that is circulating an unweighted water-base mud.

Desilters have 3-, 4-, or 5-inch (76-, 102-, or 127-millimetre) cones (fig. 81). Usually a desilter has from eight to twenty cones, depending on the amount of total fluid circulating in the well. The best location for a desilter is downstream from the shale shaker and desander.

Figure 81. Cones in a desilter (Courtesy of Derrick Corporation)

Efficiency

The amount of pressure going into the hydrocyclone and the way crew members rig it up make the most difference to its efficiency.

Pressure. The pressure of the incoming mud stream is critical to obtaining maximum efficiency with hydrocyclones. If the pressure is too low, solids do not separate well from the mud and the capacity drops. Too much pressure drastically reduces the service life of the units because the mud stream abrades the cone. Higher pressure also tends to "short circuit" the mud from feed to overflow. If the flow does not make its full trip down the cone and back up, separation cannot occur.

Large cones operate more efficiently with lower pressures than small cones. So most desanders operate at about 25 to 35 psi (172 to 241 kPa) and most desilters at 30 to 45 psi (207 to 310 kPa), measured at the inlet to the cones (not at the pump). Consult the manufacturer's operating manual for optimum conditions.

Rig Up. To work efficiently, the installers must correctly rig up the hydrocyclones. Often, the company that sells the desanders and desilters also installs them to work properly. It is important for the rig crew not to change the placement of the equipment. Or, if they suspect that rig up is a problem, they should consult the manufacturer or the manufacturer's representative.

A centrifugal pump supplies steady pressure to the hydrocyclone. Each cone size can process a specified amount of mud per minute. The installer matches the pump size to the hydrocyclone capacity by simply matching the gallons per minute (cubic metres per minute) ratings of the two. No one should use the centrifugal pump that is feeding the desander or desilter for any other purpose. As always, flood the centrifugal's suction.

What is more, installers should mount the pump upstream from any agitators in the mud tanks. Mud agitators can add air to the mud and stir up large particles that may enter the pump suction and cause problems. A screen over the pump's intake can keep out particles that could abrade the pump's impeller. The screen should have ⅝- to ¾-inch (16- to 19-millimetre) openings, which is fine enough to keep out the most damaging solids, but not so fine that it impedes the flow of mud into the pump.

Maintenance

The most common causes of internal wear to a hydrocyclone are excessive pump pressure and plugging of the apex. Prevent plugging by replacing torn shale shaker screens that allow large cuttings to enter the hydrocyclone. Also, regularly inspect and clean the suction screen to the centrifugal pump feeding the hydrocyclones.

To determine whether a desander or desilter is working properly, look at the discharge from the cone's apex. A gentle spray in the shape of a hollow cone is best, showing maximum removal of solids. A rotating semisolid spiral discharge, or "rope," means that not enough fluid is going to the underflow and solids are probably going out the overflow. The cause is that not enough mud is moving through the unit. Check that the pump is the proper size and that the pressure is correct.

Mud Cleaners

Manufacturers developed *mud cleaners* (fig. 82) in the early 1970s. They responded to a need for equipment that would remove drilled solids from weighted muds without also removing the barite and too much fluid. Mud cleaners combine desilting hydrocyclones and a very fine-mesh vibrating screen. The hydrocyclones and fine-mesh screen remove drilled solids. Mud additives and liquids return to the circulating system.

Figure 82. Mud cleaner (Courtesy of Derrick Corporation)

Separation Process

After the drilling fluid has passed through a shale shaker to remove the large cuttings, a centrifugal pump moves the mud into the hydrocyclones on the mud cleaner (fig. 83). The hydrocyclones remove silt and discharge it and the mud in the overflow into the next tank downstream. The underflow with its very fine solids passes through the bottom of the cyclones onto a fine screen below.

As the screen vibrates, drilled solids larger than the screen openings cannot pass through and go to a waste pit. The solids small enough to pass through the screen, including most barite and the liquid film around them, go into the next tank downstream along with the overflow. The size of particles separated by a mud cleaner depends on the efficiency of the desilting cones and the mesh of the screen.

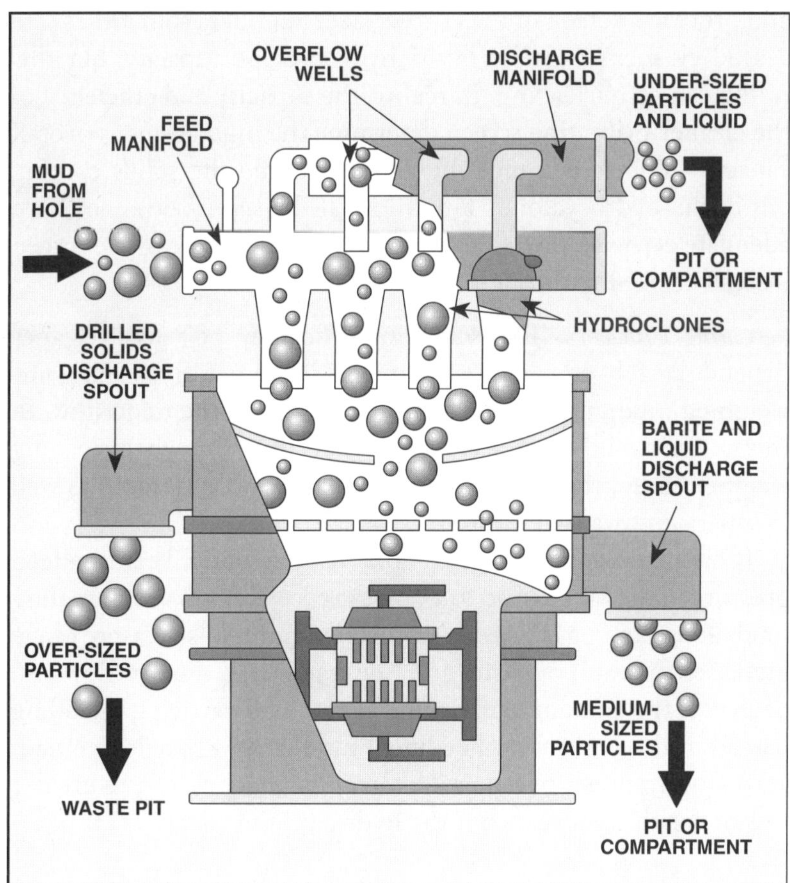

Figure 83. Mud cleaner separation process

Design of Mud Cleaners

Most mud cleaners in operation today use six, eight, ten, or twelve 4-inch (102-millimetre) hydrocyclones.

Usually, an electric motor powers a mud cleaner's vibrating screen. The vibrating screen on the cleaner removes the fine particles in the mud. A mud cleaner's screen is more fragile than a shale shaker's screen because the cleaner's screen has a finer mesh. The mesh and size of screens used on mud cleaners are important to their performance, regardless of whether the screens are round, square, or rectangular.

Capacity and Efficiency

Operators use mud cleaners mainly to remove fine solids in mud, without removing the barite. To avoid removing barite, the installer must accurately adjust the cleaner's throughput capacity and provide a screen of the correct size. Suppliers sometimes rate a mud cleaner's capacity by its hydrocyclone capacity, but this rating can be misleading. In reality, the capacity and efficiency of the cleaner's vibrating screen determine the mud cleaner's overall capacity. The screen must be able to handle the solids coming out of the hydrocyclone's underflow. If the screen does not have adequate capacity, the underflow will fall off the side of the screen instead of passing through it.

Screening Efficiency. The screening efficiency of a mud cleaner depends largely on whether the mud is weighted. In cleaning unweighted mud, most of the solids should go to the underflow. In this case the mud cleaner works like a desilter. But when cleaning weighted mud, the cleaner must salvage barite and liquid as well as separate unwanted solids.

One manufacturer suggests that rig personnel keep the feed pressure in the range of 40 to 55 psi (275 to 380 kPa) when treating muds above 12 ppg (1,438 kg/m^3). Maintaining this pressure range increases the ratio of barite and liquid going to the overflow and decreases the amount of ultrafines (very small particles, including barite) in the underflow. Feeding a mud cleaner's hydrocyclones at too high a pressure reduces screening efficiency, causes greater loss of barite, and wears out the hydrocyclone faster.

Adjusting Operation. As drilling conditions change, qualified rig personnel must carefully adjust the mud cleaner's operation to the circumstances. They determine by trial and error the exact number of cones, or manifold capacity, and the screen mesh size that will successfully handle the full circulating volume of mud. As mentioned earlier, the screen's capacity, and not the hydrocyclone's capacity, is the more important consideration. As a rule, use the finest-mesh screen possible, and vary the manifold capacity as needed by blanking off some of the cones.

Shortcuts such as using coarser-mesh screens or tolerating insufficient mud processing capacity seriously limit the effectiveness of the system. It is important for operators to recognize when they need two mud cleaners. Using one when the system needs two is false economy.

Installation

A mud cleaner sits downstream from the shale shaker and the degasser. Derrickhands often reconnect the same pump that feeds the desander or desilter to feed the mud cleaner when weighting material is added to the drilling mud. Most mud cleaners can function as desilters on unweighted mud by removing or blanking off the screen portion of the unit. In this way they may replace the desilter during top-hole drilling.

Special Applications

Operator experience and rising drilling costs have extended the use of mud cleaners into many areas other than the standard one of salvaging barite from weighted mud. Salvaging the liquid phase of a drilling mud often justifies the cost of using a mud cleaner when the liquid phase of the mud—such as oil, polymers, or water that must be trucked to the location—is expensive. The savings involved in returning most of the liquid back to the active mud system is especially significant with oil-base muds.

A mud cleaner's ability to remove most of the liquid from the drilled solids also makes it useful in areas where environmental restrictions prohibit the use of reserve pits. In such cases, the operator could rent a vacuum truck to separate the solids from the liquid in the mud tanks, an expensive choice. But a mud cleaner produces semidry solids which can be hauled off by economical dump trucks or in portable waste containers for use as landfill.

Operators can take this idea even further by using a centrifuge with a mud cleaner to form a *closed system* that eliminates discarding any whole mud (fig. 84). Closed systems are increasingly important in areas where liquid mud is an environmental contaminant. In these areas operators often have to haul the solids a considerable distance to a proper disposal site where the disposal company then charges an additional disposal fee. In a closed system, underflow from the mud cleaner screen goes to a holding tank and then to a centrifuge, which produces very fine, semidry solids. The liquid then returns to the active system. Such a system virtually eliminates the need for reserve pits, minimizes dilution, eliminates vacuum services for disposal of liquid mud, and meets environmental standards for drilling in ecologically sensitive areas, including offshore. One oil company discovered, in using such a closed system, that it saved money as well as protecting a sensitive estuary at one of its drilling sites.

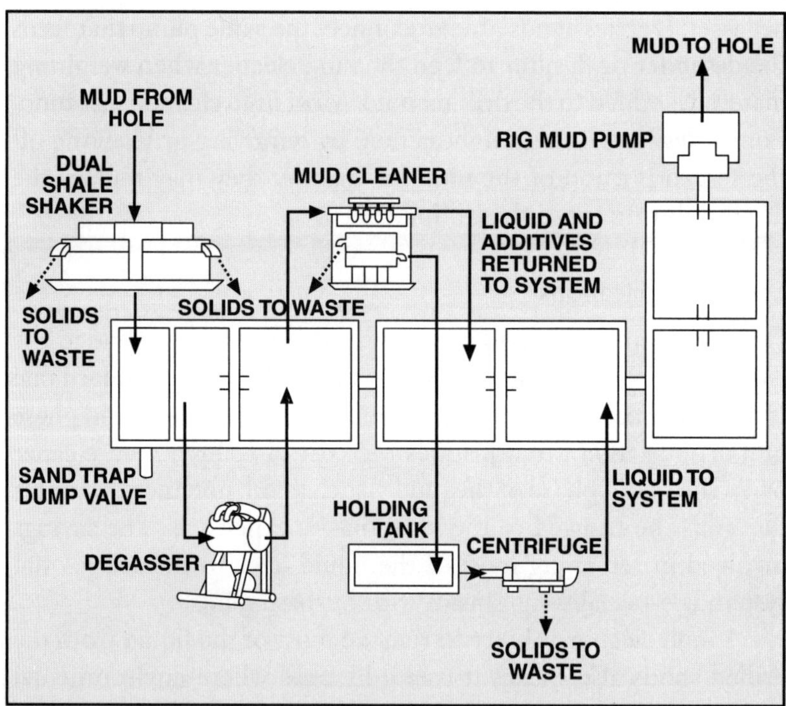

Figure 84. Closed system (unweighted mud)

Mud Centrifuges

Manufacturers offer two main types of centrifuges: the decanting centrifuge and concentric cylinder centrifuge.

Decanting Centrifuge

A *decanting centrifuge* (fig. 85) separates the fine solids that increase viscosity and recovers weighting materials. A properly operated decanting centrifuge can recover 97 to 99 percent of barite particles larger than 2 to 3 microns.

The products of a decanting centrifuge are different from those of a hydrocyclone. Instead of producing two streams of liquid, a decanting centrifuge produces a semidry stream of coarse solids and a liquid containing the smaller particles. When used to recover barite, the coarse solids contain most of the barite from the feed mud, and these return to the active system. The liquid fraction that leaves the centrifuge as overflow contains the viscosity-producing fines, which usually go to a waste pit.

The decanting centrifuge is very efficient, making a sharp cut of particles at 2 to 5 microns, depending on the specific gravity of the fluid's solids. It is very useful in many applications. For example, operators sometimes use it as a super desilter to remove drilled solids from unweighted oil-base muds. In this case, operators discard the solids as a semidry sludge and keep the valuable liquid phase.

Figure 85. Decanting centrifuge

How It Works. A decanting centrifuge has its own electric motor, centrifugal pump, and feed tank, and usually sits on a movable skid. The centrifuge itself consists of a cone-shaped steel bowl resting on its side that rotates very fast (fig. 86). Inside the bowl is a double-screw conveyor on a hollow shaft that also rotates, but at a speed slower than the bowl. A connection from the rig water supply adds water to the incoming mud stream to reduce viscosity. Generally, installers set the incoming water to supply about 3 gallons (11 litres) per minute. The diluted mud enters the centrifuge through the hollow shaft, which distributes it to the bowl. Centrifugal force created by the rotating bowl pushes the mud against the walls of the bowl. Centrifugal force also throws the larger and heavier particles against the walls of the bowl, where the conveyor blade scrapes and pushes them to the narrow end of the bowl. There they pass out through the coarse solids discharge as damp particles with no free liquid.

If coarse drilled solids and sand particles are in the mud stream entering the centrifuge, they will remain mixed with the barite returning to the system. This mixture is unavoidable because the centrifuge separates coarse and fine particles.

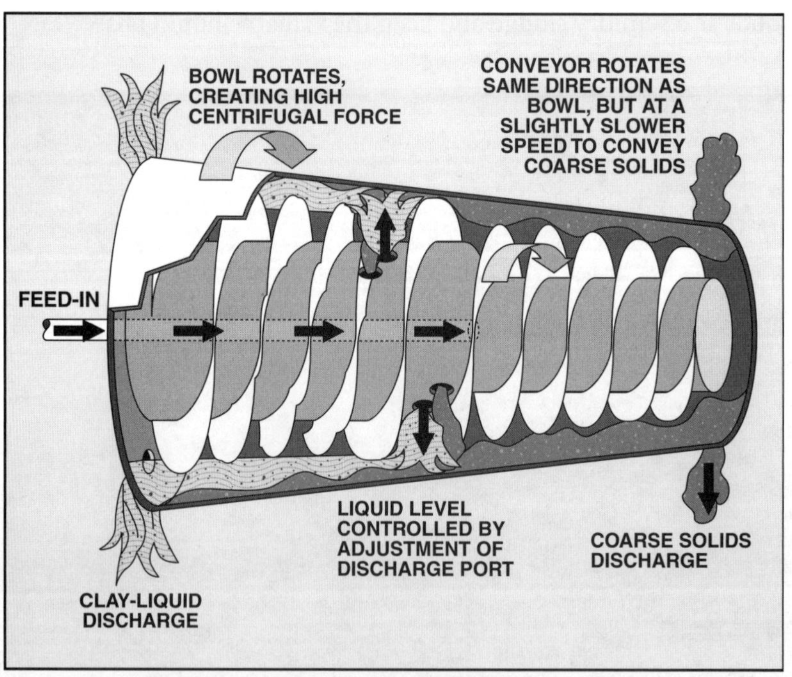

Figure 86. Operation of a decanting centrifuge

Efficiency. Normally, the crew centrifuges only a small fraction of the mud at one time. Centrifuging does not take the place of diluting a viscous mud with water—it simply permits much less dilution. If the crew centrifuges too much of the mud in the system, the process may remove too many desirable colloidal solids. This removal can make the mud have high fluid loss under high pressures and temperatures and become too thin to suspend the barite. To solve the problems associated with removing desirable solids, rig personnel can replace them with a small amount of bentonite. Add at least one sack of bentonite for each hour of centrifuge operation.

Concentric Cylinder Centrifuge

A *concentric cylinder centrifuge* for recovering barite from mud separates the mud into two liquid streams. One stream contains most of the barite, and the other contains clay, silt, chemical-bearing portions of the mud, and a small amount of barite.

How It Works. A concentric cylinder centrifuge consists of two concentric cylinders (fig. 87). The outer cylinder, or *case*, is an 8-inch (20-centimetre) pipe, 4 to 5 feet (1.2 to 1.5 metres) long. On a shaft inside this outer case is a cylinder with a slightly smaller diameter, containing a large number of perforations. The perforated cylinder, the *rotor*, revolves concentrically inside the fixed case at about 2,300 rpm.

As with a decanting centrifuge, a concentric centrifuge includes a connection to the rig's water supply to dilute the incoming mud. Usually, installers set the system to add 0.5 to 0.6 barrels of water per barrel of mud (0.5 to 0.6 cubic metres per cubic metre of mud).

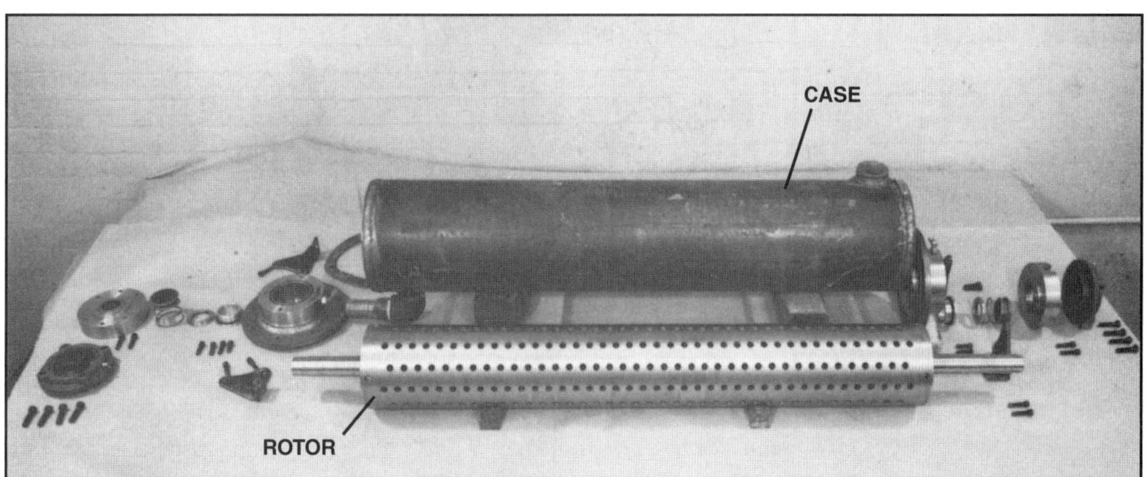

Figure 87. Operating parts of a concentric cylinder centrifuge

Two small reciprocating pumps supply mud and dilution water to the annulus between the case and the rotor. The spinning rotor causes the mud in the annulus to rotate. Centrifugal force moves the solids away from the rotor and toward the wall of the outer case. As a result, the fluid near the rotor's surface has little or no barite in it.

An underflow pump removes the barite-rich fluid along the case wall from the exit end of the separator. Manufacturers design the unit so that the discharge rate is always lower than the rate at which the feed pump moves fluid into the centrifuge. The imbalance in flow causes the barite-depleted fluid along the rotor's surface to pass through the holes in the rotor and out of the separator through one end.

Efficiency. The separation efficiency of a concentric cylinder centrifuge depends on the amount of mud the unit handles. In the range of 10 to 20 gallons per minute (2.25 to 4.5 cubic metres per hour), the particle size of the separated barite varies from 2.8 to 2.9 microns. Barite recovery ranges from 70 percent to 95 percent.

Uses of a Centrifuge

The operator may use centrifuges in the mud system for many different purposes.

Maintaining minimum solids content in weighted muds. In solids control of weighted muds, the object is to discard particles larger than barite by screening, and particles smaller than barite by centrifuging (fig. 88). Barite, and drilled solids that are the same size as barite, remain in the mud.

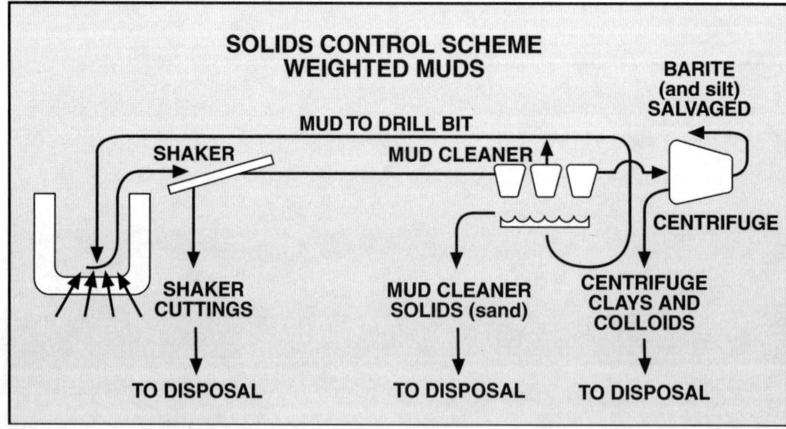

Figure 88. System to salvage barite from weighted mud. Shakers and mud cleaners remove cuttings and sand. Centrifuges salvage barite and low-gravity silts back to the active mud system. This system also discards clay solids that can cause problems.

Personnel should install the centrifuge over an active mud tank that is downstream from the point where the centrifuge pump picks up feed mud. Also, they should provide a mud agitator in the tank into which the recovered barite flows. The agitator helps ensure that the barite stays suspended in the mud tank as the mud returns to the active system. If necessary, the crew may also add bentonite, water, and new barite to the same tank.

Recovering Barite from Stored Mud. If the rig has heavy mud that is no longer in use and is stored in tanks, the operator can save money by recovering and reusing the barite in the stored mud. In this application, the centrifuge takes barite-laden mud from the storage tanks, recovers the barite, and puts it into the active drilling mud system. The operator discards most of the light and fine solids and the liquid from the stored mud. After adding the recycled barite, crew members can add water and other materials to adjust the mud properties as required.

Removing Water from Weighted Mud. When crew members add water to the mud for various purposes, the additional water increases the mud's volume and decreases the mud's weight. Operators can use a centrifuge to remove the water, which decreases the mud's volume and restores the mud's weight. The centrifuge's barite-containing stream returns to the active system. The excess liquid goes to the reserve pit where the operator discards it.

Removing Drilled Solids from Unweighted Muds. If a rig does not have a desander or a desilter, but does have a centrifuge, the centrifuge can remove coarse drilled solids, and most sizes of silt, from unweighted mud. For this purpose, the crew turns off the water supply to the centrifuge. They do not need to dilute the mud as it enters the centrifuge; they add water only as the mud system needs it. The liquid stream with fine particles returns to the active system, and the coarse drilled solids go to the waste pit. One drawback to using centrifuges for solids removal, instead of desanders and desilters, is that centrifuges usually cannot handle the full circulating capacity of the mud system. Desanders and desilters can handle high volumes of fluid with no difficulty.

If the rig does have desanders and desilters, the operator can also use a centrifuge to remove solids from unweighted mud. In this case, the shaker removes the cuttings, while the desander and desilter remove sand- and silt-sized solids. Finally, the centrifuge removes the very fine silts (fig. 89). The clean mud returns to the active system.

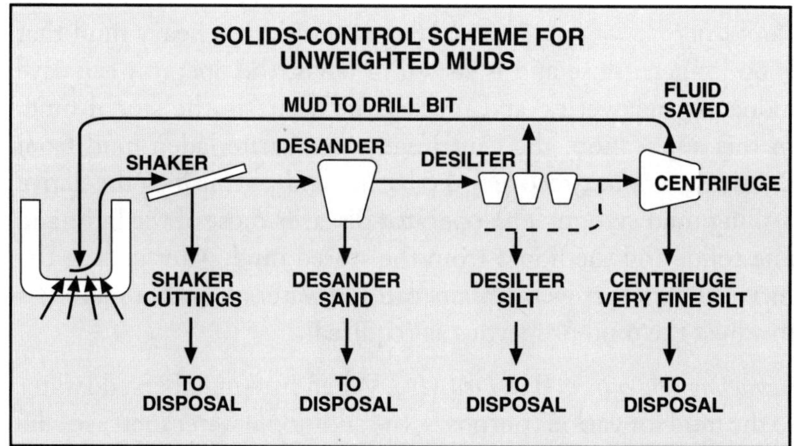

Figure 89. System to remove solids from unweighted mud. Shakers, desanders, and desilters remove cuttings, sand, and silt. Centrifuge removes very fine silts.

Maintaining Weight of Oil Muds. Operators can use a centrifuge to maintain minimum weight in invert-emulsion or oil-base muds. Centrifuges are very good at removing solids of a specific size and they lose very little of the liquid phase. The disadvantage of using a centrifuge for this purpose is its relatively small capacity.

Processing the Underflow from Mud Cleaners. Operators often use centrifuges to process the underflow from a mud cleaner. The low-volume capacity of the centrifuge is not a problem in this application because the mud cleaner's screen also has low-volume capacity. Since the two capacities match, the underflow from the cleaner does not overwhelm the centrifuge. Also, the mud cleaner's screen prevents coarse solids from entering the centrifuge, which reduces centrifuge maintenance costs and downtime.

To summarize—

- Solids control equipment controls the amount, size, and type of solids in drilling mud.

Desanders and desilters

- Adjust the pressure of the mud stream entering desanders and desilters for the best separating efficiency.
- Inspect and clean the feed pump's suction screen regularly.
- Check the underflow—it should be in the form of a spray, not a rotating rope.

Mud cleaners

- Maintain correct pump pressure, number of cones, and size of screen mesh for best separating efficiency.

Centrifuges

- Watch the mud's viscosity when using a centrifuge, and replace the desirable fines that have been discarded with a small amount of bentonite.
- Uses of a centrifuge
  - Recovering barite from weighted mud
  - Recovering barite from stored mud
  - Removing water from weighted mud
  - Removing drilled solids from unweighted muds
  - Maintaining weight of oil muds
  - Processing the underflow from mud cleaners

Mud-Gas Separators and Degassers

A mud-gas separator is a relatively small closed vessel or tank, usually made from steel plate (fig. 90). Rig personnel send mud and gas to it when the well is closed in on a gas kick. It can remove large quantities of gas from mud.

Figure 90. Mud-gas separator (Courtesy of Nabors Drilling USA)

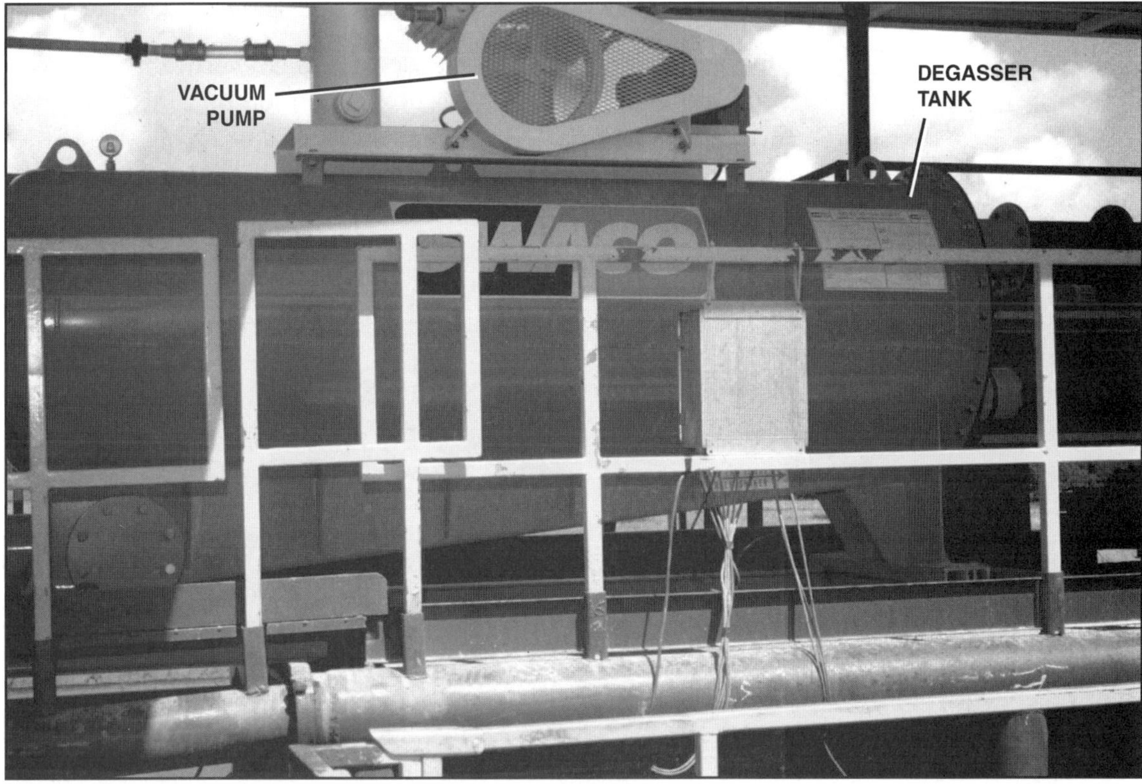

Figure 91. Vacuum degasser (Courtesy of Nabors Drilling USA)

A *vacuum degasser* is also a steel tank, or vessel, but unlike a mud-gas separator it removes entrained gas from mud being normally circulated through the open mud tanks (fig. 91).

Mud-Gas Separators

Crew members often call a mud-gas separator a *gas buster*. Its primary purpose is to remove gas from mud when rig personnel are circulating a gas kick and mud out of a closed-in well. In this situation, mud does not flow out of the bell nipple and into the return line to the mud tanks. Instead, personnel shut in the top of the well with special valves called the blowout preventers. Shutting in the well requires them to circulate mud and gas from the kick out through a special valve (the *choke*) and special piping (*choke manifold*). The choke allows personnel to hold back-pressure on the well to prevent more kick fluids from entering. At the same time, it allows them to pump the kick fluids (gas) and mud to the choke manifold. Installed downstream from the choke manifold, the mud-gas separator receives the gas and mud and separates them.

How It Works

Mud-gas separators are simple devices. It is a vertical vessel with a series of baffles inside that spread the mud out, increasing its surface area (fig. 92). As the mud splashes down from one baffle to the next, the gas breaks out and flows up through a center vent. A vent line at the top carries the gas far enough from the rig to safely flare it.

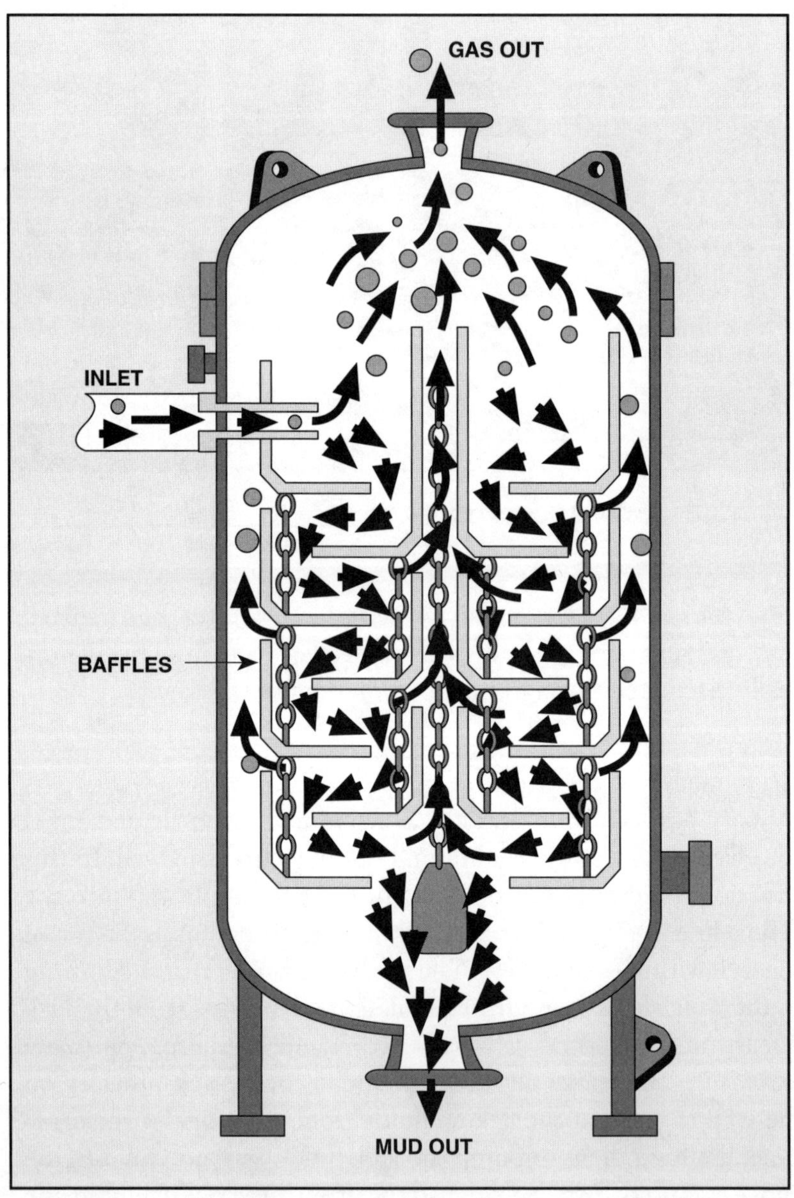

Figure 92. Operation of a mud-gas separator

Gas-free mud flows out near the bottom, then upward through a vertical pipe (a *riser*), and finally through a horizontal line to the shale shaker and mud tanks. A float valve keeps the fluid level in the separator above the pipe that returns mud to the tank. If it falls below the pipe, gas will enter the mud line instead of going out the vent to be flared.

When circulating through the choke and manifold, the driller can keep moderate pressure, or back-pressure, in the separator without hindering its degassing ability. Back-pressure is hydrostatic pressure plus pumping pressure. Keeping back-pressure on the mud-gas separator is good practice. Without some pressure on the gas as it enters the vessel, the gas would expand quickly. Large volumes of expanding gas entering the separator can blow it right off its base. Also, back-pressure reduces the rate of gas expansion in the mud stream and moderates surges as slugs of gas and mud reach the surface. Working pressure of the unit may be as much as 100 psi (690 kPa).

Vacuum Degassers

Gas-cut mud is drilling mud that contains entrained formation gas. What happens is that the bit enters a formation that contains a small amount of gas. At the bottom of the hole, where the hydrostatic pressure is high, the gas stays in solution in the mud. As the gas and mud circulate to the surface, however, hydrostatic pressure falls. As a result, the gas comes out of the mud at the surface and appears as bubbles or froth in the mud in the tanks. Gas-cut mud circulated from the bottom may be several pounds per gallon (several hundred kilograms per cubic metre) lighter than the fluid pumped into the hole. It is vital for crew members to remove the gas from the mud before it circulates back down the hole. For one thing, gas in the mud reduces the mud pump's efficiency. Worse, however, is that the lightweight, gas-cut mud could allow a kick to occur. Gas-cut mud may not develop enough hydrostatic pressure to keep formation pressure in check. The result is a kick.

Running the return mud across the vibrating screen of the shale shaker, using settling action in the tanks, and stirring the mud release some of the entrained gas from drilling mud, but may not release it all. For this reason some wells require a *degasser* to treat the mud. The degasser separates gas and vents it at a safe distance from the well and returns the gas-free mud to the mud tanks.

A vacuum degasser (fig. 93) uses a vacuum pump to help draw the gas from the mud. One popular type is installed between two mud tanks. It takes the gas-cut mud from one tank and releases the degassed mud into the second tank. Gas-cut mud enters the degasser tank through a suction pipe. Mud enters the pipe because the vacuum pump on the degasser lowers the pressure (creates a *vacuum*) inside the degasser tank. Mud enters the degasser through a feed pipe near the top of the vessel. Inside the vessel, the feed pipe, which is closed at its far end, has its top cut off so that the mud can spill over the sides onto a baffle that extends the full length of the feed pipe and slopes downward.

As the mud streams down the sloping baffle, the vacuum pulls the gases out of the mud. The degassed mud flows to the bottom of the vessel to a discharge pipe, which is a downspout into the second mud tank.

Because the vacuum in the degasser is strong enough to prevent the mud from flowing out, the manufacturer installs a hydraulically operated jet in the downspout. A centrifugal pump picks up mud from one of the downstream mud tanks and sends it to the jet. The fast-moving jet of mud creates an area of pressure that is lower than the pressure in the degasser vessel. The mud can therefore flow from the degasser.

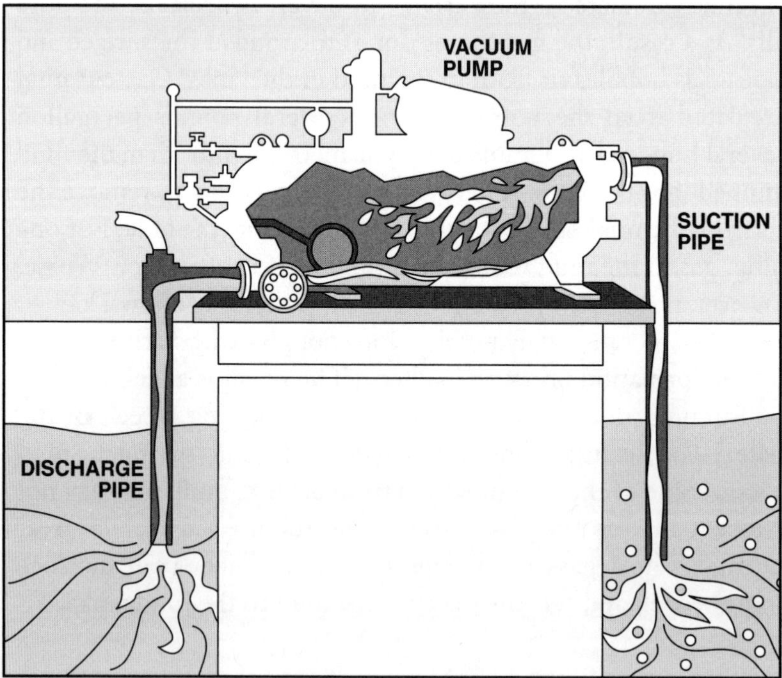

Figure 93. Operation of one type of vacuum degasser

The manufacturer also installs safety devices to prevent mud from entering the vacuum pump. Further, the unit has washdown lines to permit cleaning the machine so that it does not become plugged with cuttings, settlings, and other materials. This type of degasser can handle mud containing lost circulation materials, such as mica, nut shells, and so on, at circulation rates exceeding 600 gpm (2.27 m^3/min). However, it works better on thin muds than it does on thick muds because gas cannot break out of thick mud as easily.

Centrifugal Degasser

A *centrifugal degasser* uses a vacuum pump to draw the mud into two degassing tubes through an intake manifold (fig. 94). Once the mud is inside, the tubes spin, and centrifugal force disperses the mud into thin layers around the walls of the tubes. The vacuum then pulls the gas out from inside an expansion tank. Also, a sprayer sprays the mud onto a baffle in the tank, which breaks out any remaining gas. A blower (not visible in the figure) discharges the gas to a flare line. A special centrifugal pump returns the degassed mud to the mud system.

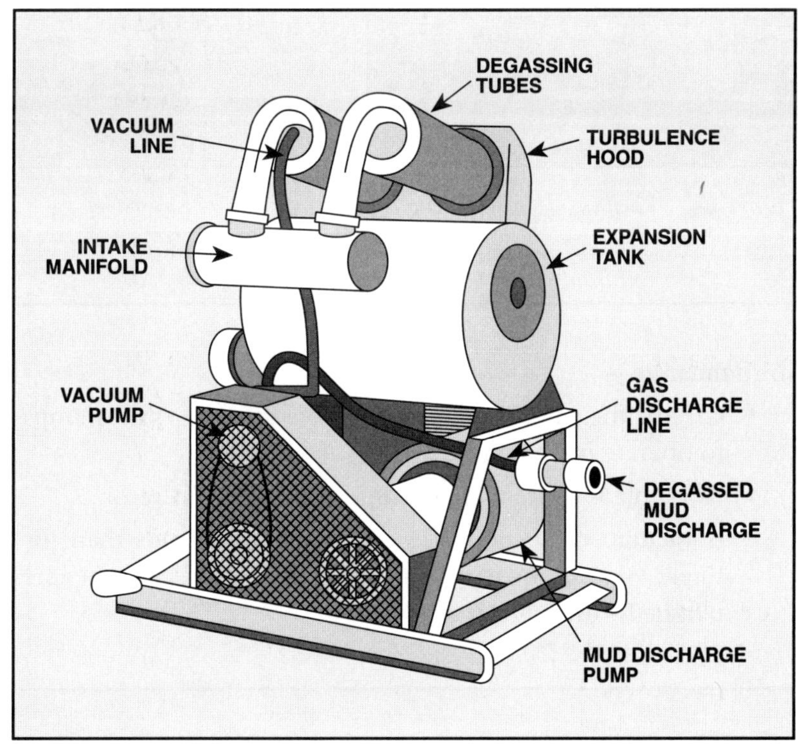

Figure 94. Centrifugal degasser

Eductor Design

Another type of degasser uses an *eductor*, which is a specially shaped tube that reduces pressure as it increases velocity of mud flowing through it (fig. 95). The reduced pressure allows the gas to break out of the mud. A centrifugal pump powers the eductor. Both mud and separated gas leave the degasser chamber in one stream. The gas exits to a flare. The degassed mud falls into a mud tank. Gas, once broken out of solution in the mud, does not redissolve.

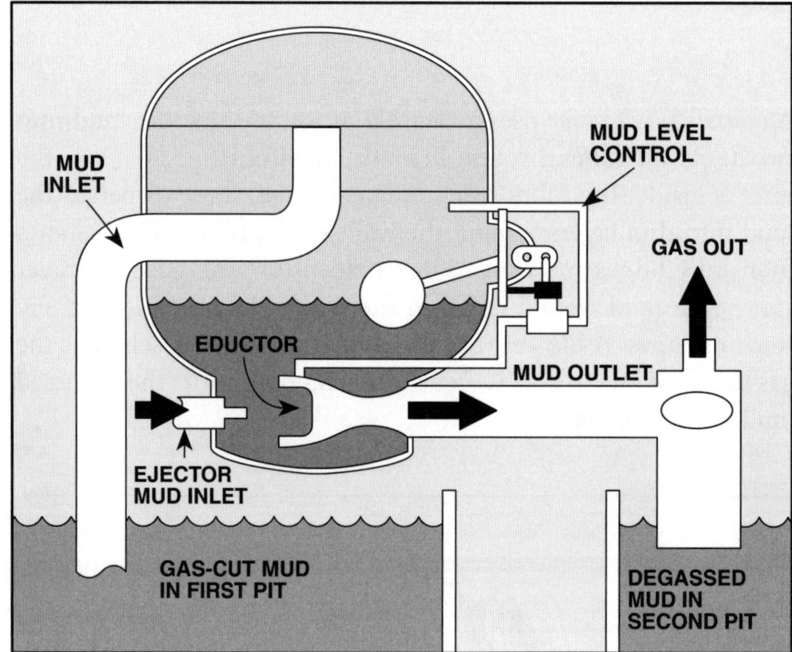

Figure 95. Vacuum degasser using an eductor

To summarize—

- Circulating gas-cut mud is dangerous and causes the pumps to operate inefficiently.
- The hole is shut in when using a mud-gas separator.
- A vacuum degasser works better for thin muds than for thick. Wash it down periodically so that it does not become plugged with solids in the mud.

Other Equipment

Other equipment in the circulation system includes mixers and agitators to mix the mud and its additives thoroughly, and instruments to monitor the volume and weight of mud in the system, its flow rate, and its temperature.

Mud Mixers and Agitators

Mixing the huge quantities of water or oil with additives to formulate a drilling mud requires heavy-duty equipment. The crew often adds solids and chemicals through a hopper inside of which is a high-velocity jet. The jet is an area of fast-moving liquid at a low pressure. The speed and low pressure ensure that the materials entering the stream are thoroughly mixed into it. Operators install agitators on the mud tanks to stir the mud so that as it passes from one tank to another, heavy additives stay in suspension.

Jet Hoppers

Drilling contractors have used *jet mixing hoppers* for more than 50 years, and universally use to add dry or liquid materials to liquid mud. The original developers intended them for mixing cement and water into a slurry for oilwell cementing. The hoppers worked so well that contractors soon adopted them for mud mixing. Both high-pressure and low-pressure units are available. Either type ensures efficient, quick distribution of solids into the mud.

How It Works. The hopper is funnel-shaped with an opening at the bottom. On one side of the opening is a jet. Immediately downstream from the jet is a venturi tube (fig. 96). During operation, a centrifugal pump moves mud from an active mud tank through the jet and venturi tube and back to a second tank. *Venturi tubes* are a type of eductor. They work on the principle that a fluid flowing through a constriction has increased velocity and reduced pressure.

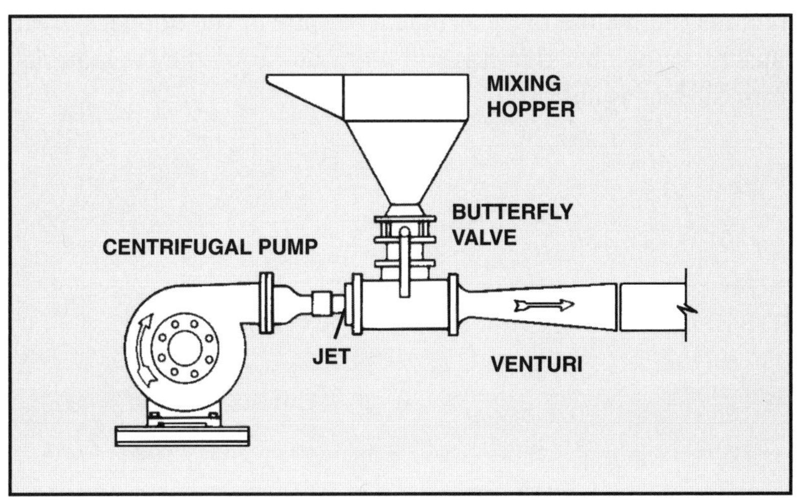

Figure 96. Centrifugal pump and jet hopper for mud mixing

The mud passes through the tube so fast that it lowers the pressure at the hopper's opening to below atmospheric pressure, which creates a vacuum. The vacuum sucks the dry material in the hopper into the stream of mud, where it mixes with the mud. The jet just upstream from the venturi tube increases the vacuum even more.

The dry material can be a powder, such as clay, bentonite, barite, and chemicals, or a pulverized material, such as cellophane and nut hulls. (Remember, however, that it is not safe for crew members to mix caustic soda into the mud with a jet hopper. It is too easy for the caustic to splash back into the face and body of the person adding it.)

Some models have a butterfly valve between the hopper and the jet nozzle. During periods when mud is flowing through the jet but crew members are not adding materials to it, air can get into the mud stream. To prevent air from mixing with the mud passing through the jet while not adding materials, a crew member can close the valve to keep air out.

Crew members can use an alternate method to add liquids to the mud. In the alternate method, they do not add the liquid through the hopper's funnel. Instead, they rig up a hose with a valve on it between the container of liquid and the hopper's jet. When a crew member opens the valve on the hose, the jet pulls the liquid into the mud stream. This method is very effective for emulsifying oil or adding materials that have been premixed with fresh water to a brine system.

Capacity. The capacity of a hopper is related to the size of the jet and the volume of fluid pumped through the jet. Even a small jet hopper can satisfactorily mix 200 to 400 pounds (90 to 180 kilograms) of weighting material per minute into the mud stream. High-capacity units can handle twice as much. In spite of the hopper's great capacity, remember that it is still important to shake the material from the bag as uniformly as possible.

Mud Agitators

Mud agitators stir the mud while it is in the mud tanks. They are necessary for several reasons. The first is to keep the mud phases mixed during the slow travel of mud through the tanks, sometimes less than 2 feet (0.6 metre) per minute. Weighting material, for example, can drop out of suspension. Another is to break up muds that have high gel strength. Recall that drilling mud often contains additives that cause it to gel when it is not moving, so that the cuttings and other solids remain distributed throughout the mud during trips. Such muds gel in the tanks unless they are constantly stirred. A third reason is to allow gas bubbles to escape in minor gas cutting. A final purpose is to help mix the mud when adding materials to the system through a jet hopper.

A mud agitator is basically a propeller or paddle powered by an electric motor (fig. 97).

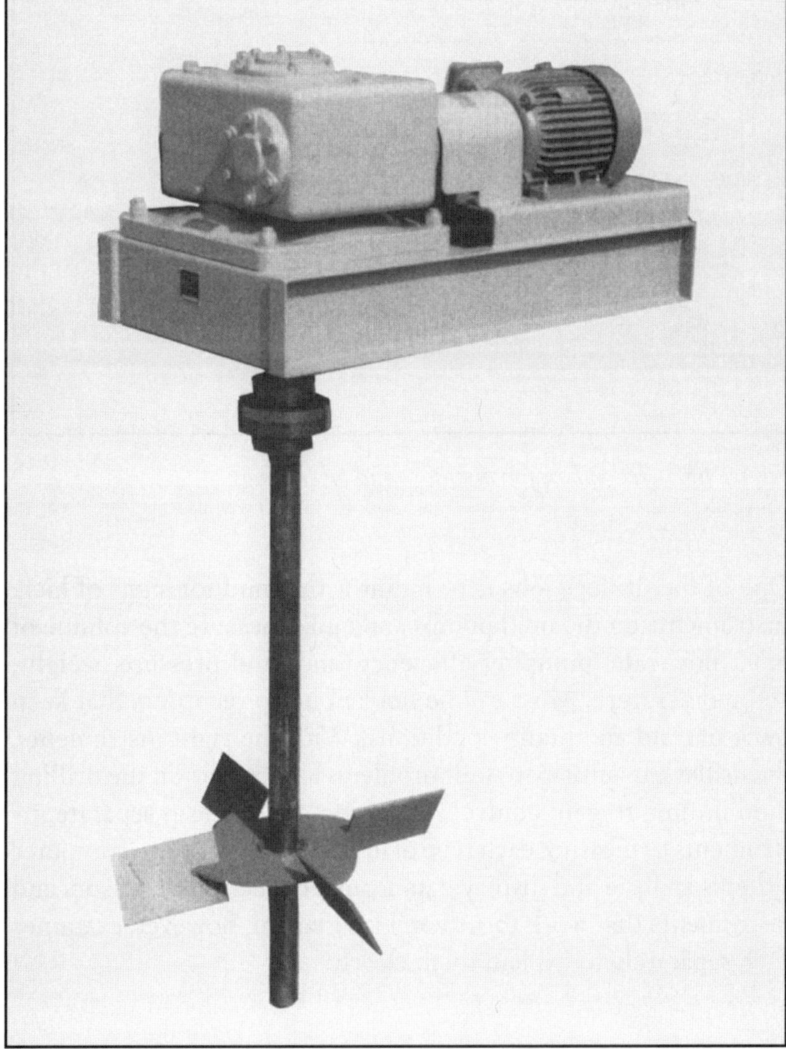

Figure 97. Mud agitator

Jet Siphons

A *jet siphon* transfers mud, sand, or cuttings from a mud tank to the reserve pit or to a special tank for disposal. It consists of a 1- to 2-inch (25- to 50-millimetre) pipe that goes near the bottom of the mud tank, turns back up, and ends inside the mouth of a larger pipe (fig. 98). To create the siphoning effect, a centrifugal pump pumps drilling mud or water into the small pipe. Its velocity creates an area of low pressure that pulls mud from the tank.

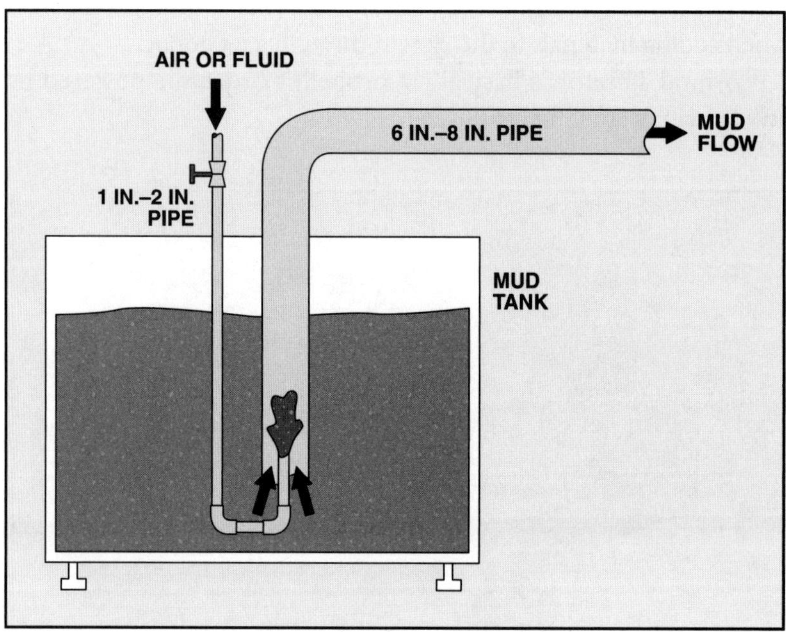

Figure 98. Jet siphon for moving mud or cuttings

Mud Monitoring Instruments

One of the driller's jobs is to monitor the mud for signs of kick. Instruments on the mud pumps and tanks measure the volume of mud, flow rate, pumping efficiency, and mud pressure, weight, and temperature. Most can be hooked up to recorders that keep track of mud and pump conditions. With the right instruments, the driller can anticipate well problems and condition the drilling fluid in time to gain control. Older technology uses separate instruments to measure each type of information. Modern computer systems analyze and display data from a collection of sensors and instruments that work together. This manual, however, examines each separately to explain them clearly.

Mud Volume

A *pit-level indicator*, or *mud volume totalizer*, shows the level of mud in the mud tanks.

How It Works. A pit-level indicator can be pneumatic or electronic. Both types use several floats that move up or down with the level of mud in the tank. Electronic sensing elements or compressed air devices on the floats send signals to a special relay that averages the readings. The relay then transmits the average to an indicator on the rig floor (fig. 99). The mud volume totalizer may include a recorder to keep track of total volume and changes over time.

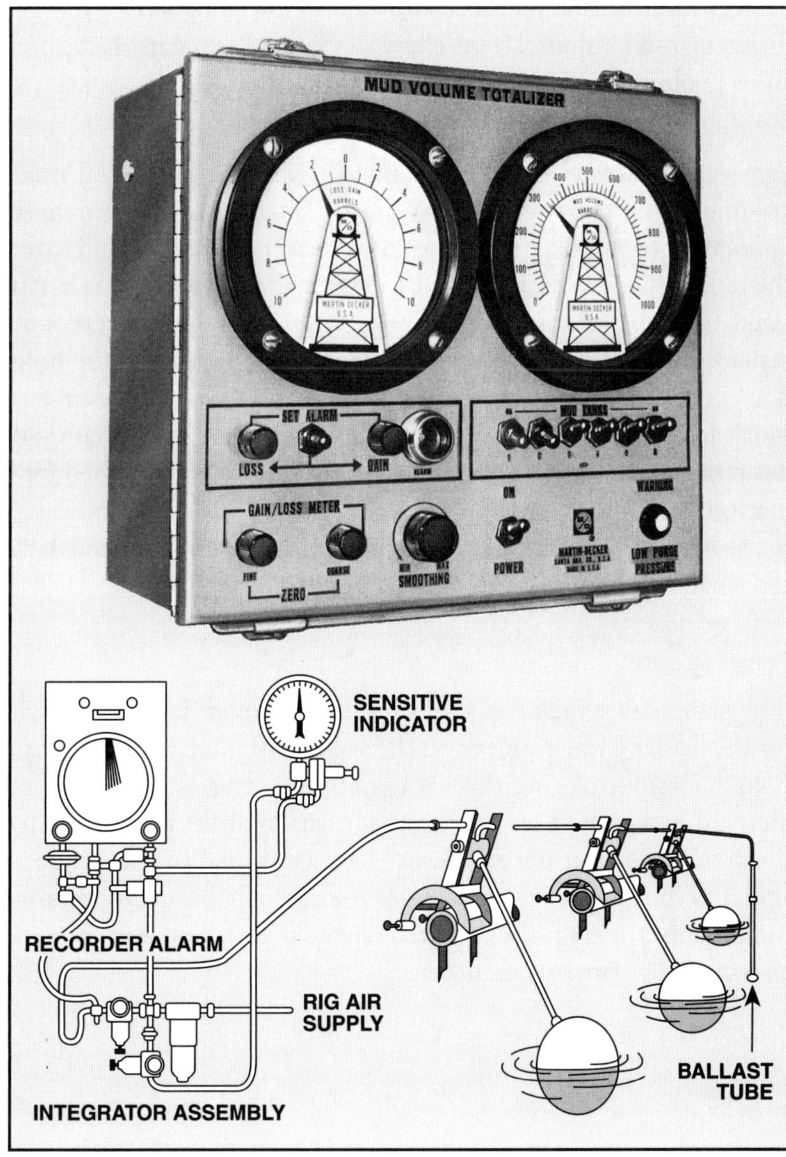

Figure 99. Arrangement of a pit level indicating instrument (Courtesy of M-D Totco)

The mud volume totalizer can trigger an alarm for high level or low level in the tanks. The driller can set the alarm to go off when the level changes by only a few barrels (a fraction of a cubic metre). The sensitive indicator shows variations of as little as one barrel (0.16 cubic metre).

Purpose. Mud volume totalizers are especially valuable in warning of kicks that the crew swabs in when tripping out of the hole. The mud level in the trip tank should fall as the mud replaces the space in the hole left by the pipe. If the pumping rate has not increased and the mud level rises instead of falling, the driller knows that crew members have probably swabbed formation fluid into the hole.

Pit-level indicators are also useful in checking slow gains or losses in mud volume. The chart made by a standard indicator shows volumes over a 24-hour period so the driller can keep a constant watch on what is happening downhole.

Trip Tank Indicator. The derrickhand monitors changes in mud volume in the trip tank using a similar instrument, the *trip tank indicator*. This tool lets the crew know to add or remove mud from the tank if the level is too low or too high. Crew members use trip tanks to keep accurate track of how much mud the hole takes to replace the drill stem as they remove it from the hole. If the hole does not take the right amount of mud (either too little or too much to replace the drill stem volume), then they know that mud has either been lost to a formation or that they have swabbed formation fluid into the hole. Either too much or too little indicates that something is wrong and that crew members should check to see what is going on.

Pump Speed

The *pump stroke indicator*, or *pump stroke counter*, is another way for the driller to keep an eye on the volume of mud downhole (fig. 100). A pump stroke indicator is a small generator. When installers attach it to any shaft on a pump that turns in direct proportion to the pump speed, it generates an electrical signal that is proportional to the number of strokes per minute. The signal powers an indicator (fig. 101) at the driller's console that shows spm and may also send the data to a recorder.

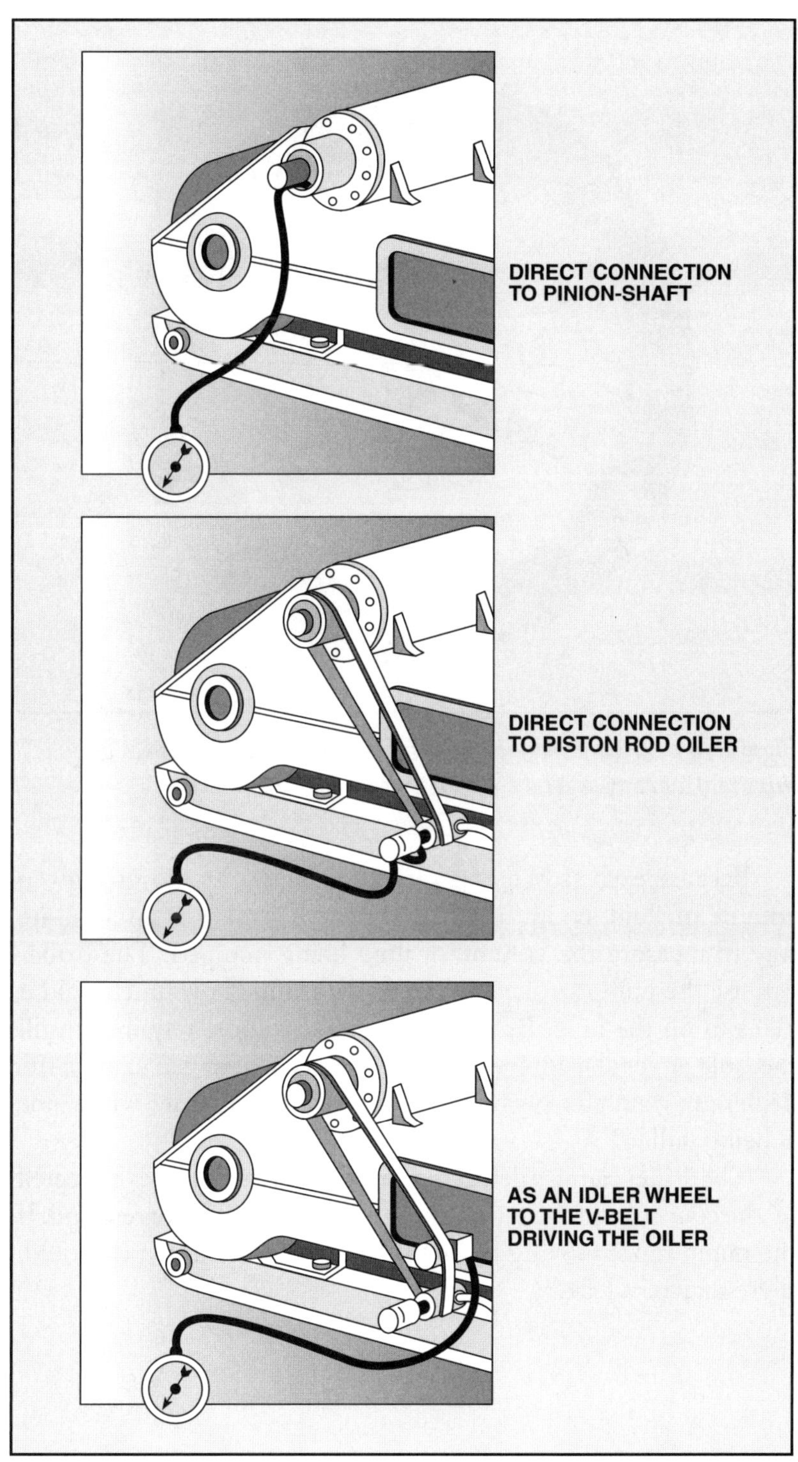

Figure 100. A pump speed indicator can be attached to the mud pump in several ways.

Figure 101. Pump speed indicator, calibrated to show strokes per minute (Courtesy of M-D Totco)

Because each stroke of a reciprocating pump moves a given volume of fluid, counting the number of strokes is an accurate way to measure the volume of fluid being pumped. The driller can use the pump stroke counter to determine how much fluid it takes to fill the hole. By noting the mud volume required to fill the hole at regular intervals, usually every five or ten stands, the driller can watch for swabbing that may be taking place while pipe is being pulled.

The driller can also detect pump trouble if the volumes indicated by the counter and the mud volume totalizer do not correspond. If the pump is not moving the actual volume of mud that it should, there may be a leak.

Pump Pressure

Pump trouble, pipe trouble, and downhole trouble all show up initially as a change in mud pump pressure. Drillers watch the pump stroke indicator and mud pump pressure gauge (fig. 102) to get the complete picture. To understand the meaning of the pressure gauge reading, the driller must also know whether the pump speed has changed.

If the pump pressure is too low, the problem could be that the drill string has a loose joint or a break, the bit nozzles are washed out, or circulation is being lost to the formation. Low pressure in combination with high spm (pump speed) could indicate that the pump is not working efficiently—liners or packings may need replacing. It could also indicate a kick. The added pressure from the formation helps the mud move up the annulus, so the pump does not have to work as hard. A pressure that is too high could mean that the mud's density or viscosity has increased.

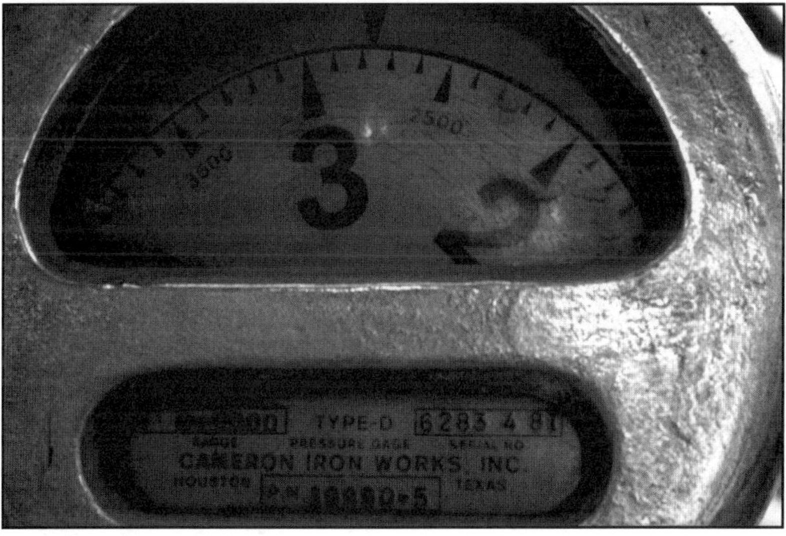

Figure 102. Mud pump pressure gauge (Courtesy of Nabors Drilling USA)

Flow Rate

Knowing the flow rate gives drillers another piece of information that helps them "see downhole."

Flow Sensor. A *flow sensor* (fig. 103) senses and indicates the rate of mud flow from the return line of the well. The key elements are a paddle that sits in the return flow line, which is connected to a torque indicator. The returning mud deflects the paddle faster or slower, and the torque indicator measures the turning force that the mud exerts on the paddle. The system electrically or pneumatically transmits a signal to an instrument on the rig floor. The flow sensor works independently of level changes in the mud tanks and is very accurate.

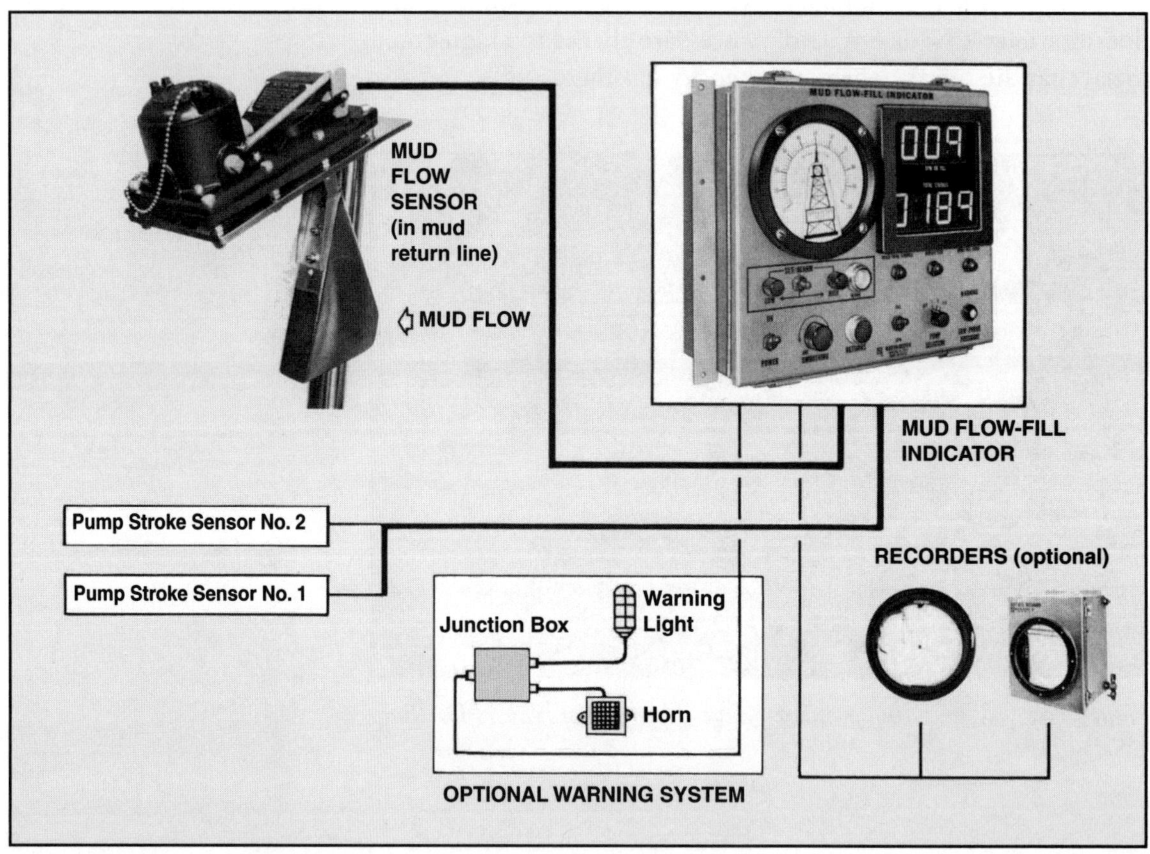

Figure 103. A return-flow sensing system (Courtesy of M-D Totco)

If the sensor indicates an increase in the rate of the mud returning from the hole, it may be a sign of a kick. If the driller has not increased the pump's speed and the flow sensor indicates an increase in the mud-return rate, then it may be that formation fluids are pushing mud out. To check for a kick, the driller can do a flow check. A flow check occurs when the driller stops the mud pump, waits for a short period, and checks to see if mud flows out of the return line even though the pump is shut down. If mud is flowing, then it is likely that kick fluids are forcing the mud out of the annulus and the driller should initiate shut-in procedures.

Differential Flowmeter. A *differential flowmeter* measures the difference between the rate of mud flow going in and coming out of the hole. It uses an electronic inflow sensor in the mud pump's discharge line and an electronic outflow sensor in the mud return line at the wellhead. The signals from the sensors measure input and outflow rates, electronically compare them, and send the difference to a strip chart recorder. When the flow rates are equal, the chart shows zero (zero is usually in the center of the chart). If the mud outflow is greater than inflow, the recorder shows the difference on one side of the chart as a gain. If inflow is greater than outflow, it shows the difference on the opposite side of the chart as a loss.

A mud flowmeter is quite sensitive and responds immediately to a change in flow rate as small as 5 gallons per minute (20 litres, or 0.02 cubic metres, per minute). It measures only changes that occur downhole. Surface operations such as mud transfer, switching pumps, and mud dilution or treating do not affect either sensor. If the driller notices that the mud outflow is greater than inflow, it may be a sign of a kick.

Mud Weight and Temperature

A *mud weight indicator*, or a *mud-density device*, also aids the driller in recognizing a kick. Like the flowmeter, a mud weight indicator has two sensors. The mud-in sensor in the suction tank measures the density, or weight, of the mud before it goes downhole. The mud-out sensor is in the shale shaker box or the mud return line. It detects any density changes in the mud as it returns. The sensors may be pneumatic (fig. 104) or electric (fig. 105).

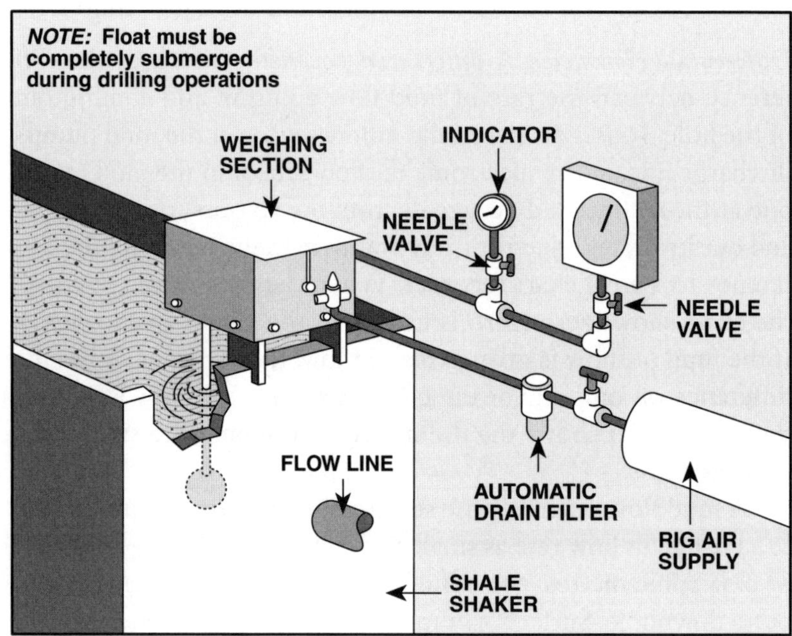

Figure 104. Air-actuated mud density recording device

The mud weight indicator may have a second set of sensors that measure the temperature of the mud before it goes into the hole and after it comes out of the hole. A change in temperature during drilling can indicate that formation fluids are entering the hole or that some part of the system has washed out. If gas enters the hole from a formation, it expands slightly because the hydrostatic pressure in the hole is less than the pressure in the formation. Expanding gas drops in temperature. The temperature drop may be small but the sensitive temperature sensor on the device is able to detect it.

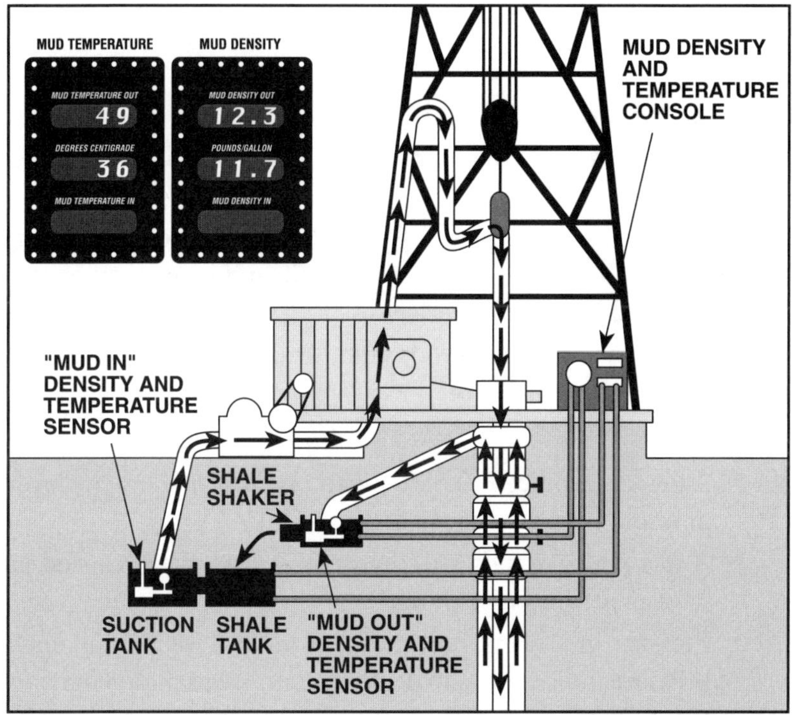

Figure 105. Electrical mud density and temperature recording device

If a washout has occurred, the return mud will not be as hot because a small amount of mud is not being circulated all the way to the bottom of the hole, where it is hottest. Instead, it is going out the washout in the pipe at some point above the bottom, where it is not as hot. The instrument's temperature sensors are very sensitive and can detect very small changes in temperature.

To tell the difference between a washout and gas kick, the driller can run a flow check. If the well flows with the pump off, then the chances are that the borehole has encountered a gas kick.

To summarize—

Mixers and agitators

- Jet hoppers are used to mix dry materials or liquids into the drilling mud.
- Mud agitators keep the dispersed solids in the mud mixed, break muds with a high gel strength, and help mix the mud when adding mix materials.
- Jet siphons transfer mud, sand, or cuttings from the mud tank to the reserve pit.

Instrumentation

- A pit-level indicator and trip tank indicator show the level of mud in the mud tanks and trip tank.
- A pump stroke indicator counts the mud pump's speed in strokes per minute (spm).
- Pressure indicators, flow rate sensors, and weight and temperature indicators help drillers monitor circulation, giving them "eyes" downhole.

Glossary

A

active mud tank *n*: one of usually two, three, or more mud tanks that holds drilling mud that is being circulated into a borehole during drilling. They are called active tanks because they hold mud that is currently being circulated.

additive *n*: 1. in general, a substance added in small amounts to a larger amount of another substance to change some characteristic of the latter. In the oil industry, additives are used in lubricating oil, fuel, drilling mud, and cement. 2. in cementing, a substance added to cement to change its characteristics to satisfy specific conditions in the well. A cement additive may work as an accelerator, retarder, dispersant, or other reactant.

aerated mud *n*: drilling mud into which air or gas is injected. Aeration with air or gas reduces the density of the mud and allows for faster drilling rates. The lighter aerated mud does not develop as much pressure on bottom as a normal mud. The lower pressure allows the cuttings made by the bit to easily break away from the bit's cutters; the cutters therefore always contact fresh, undrilled formation.

agitator *n*: a motor-driven paddle or blade used to mix the liquids and solids in drilling mud.

air bazooka *n*: a special aeration unit that forces air into dry mud material (such as bentonite), and which assists crew members in transferring the dry material from a bulk tank on the rig to a transport truck.

air drilling *n*: a method of rotary drilling that uses compressed air as the circulation medium. The conventional method of removing cuttings from the wellbore is to use a flow of water or drilling mud. Compressed air removes the cuttings with equal or greater efficiency. The rate of penetration is usually increased considerably when air drilling is used; however, a principal problem in air drilling is the penetration of formations containing water, since the entry of water into the system reduces the ability of the air to remove the cuttings.

air slide *n*: a mechanism using pressurized air through a diaphragm to fluidize powdered materials so that they will flow from a delivery truck to a storage tank on the rig site.

alkalinity *n*: the quality of being basic. The strength of a liquid's alkalinity is measured by pH; a pH above 7 is alkaline. See *pH*.

American Petroleum Institute (API) *n*: oil trade organization (founded in 1920) that is the leading standardizing organization for oilfield drilling and producing equipment. It maintains departments of transportation, refining, marketing, and production in Washington, D.C. It offers publications regarding standards, recommended practices, and bulletins. Address: 1220 L Street, NW; Washington, DC 20005; (202) 682-8000.

amine salt *n*: an organic compound derived from ammonia, in which organic compounds replace one or more of the hydrogen atoms in the ammonia.

anhydrite *n*: the common name for anhydrous calcium sulfate, $CaSO_4$.

annular space *n*: the space between two concentric circles. In the petroleum industry, it is usually the space surrounding a pipe in the wellbore; sometimes termed the annulus.

annular velocity *n*: the rate at which mud is traveling in the annular space of a drilling well.

annulus *n*: see *annular space*.

antidifferential sticking additive *n*: a chemical added to the drilling fluid to minimize the possibility of the drill stem becoming stuck to the side of the hole.

API *abbr*: American Petroleum Institute.

API gravity *n*: the measure of the density or gravity of liquid petroleum products in the United States; derived from relative density. API gravity is expressed in degrees, 10° API being equivalent to 1.0, the specific gravity of water. See *specific gravity*.

API Spec. 7 *n*: abbreviation for *API Specification for Rotary Drilling Equipment*, a publication that lists the specifications for drilling equipment.

apparent viscosity *n*: the viscosity of a drilling fluid as measured with a direct-indicating, or rotational, viscometer.

attapulgite *n*: a fibrous clay mineral that is a viscosity-building substance; used principally in saltwater-base drilling muds. Also called fuller's earth.

B

back-pressure *n*: 1. the pressure maintained on equipment or systems through which a fluid flows. 2. in reference to engines, a term used to describe the resistance to the flow of exhaust gas through the exhaust pipe. 3. the operating pressure level measured downstream from a measuring device.

barite *n*: barium sulfate, a mineral frequently used to increase the weight or density of drilling mud. Its specific gravity is 4.2 (i.e., it is 4.2 times denser than water). See *barium sulfate*, *mud*.

barium sulfate *n*: a chemical compound of barium, sulfur, and oxygen ($BaSO_4$), which may form a tenacious scale that is very difficult to remove. Also called barite.

barrel (bbl) *n*: 1. a measure of volume for petroleum products in the United States. One barrel is the equivalent of 42 U.S. gallons or 0.15899 cubic metres (9,702 cubic inches). One cubic metre equals 6.2897 barrels. 2. the cylindrical part of a sucker rod pump in which the pistonlike plunger moves up and down. Operating as a piston inside a cylinder, the plunger and barrel create pressure energy to lift well fluids to the surface.

baryte *n*: variation of barite. See *barite*.

bbl *abbr*: barrel.

bentonite *n*: a colloidal clay, composed primarily of montmorillonite, that swells when wet. Because of its gel-forming properties, bentonite is a major component of water-base drilling muds. See *gel*, *mud*.

bentonite extenders *n pl*: a group of polymers that can maintain or increase the viscosity of bentonite while flocculating other clay solids in the mud. With bentonite extenders, desired viscosity can often be maintained using only half the amount of bentonite that would otherwise be required.

bent sub *n*: a short cylindrical device installed in the drill stem between the bottommost drill collar and a downhole motor. Its purpose is to deflect the downhole motor off vertical to drill a directional hole. See *drill stem*.

BHP *abbr*: bottomhole pressure.

BHT *abbr*: bottomhole temperature.

bit *n*: the cutting or boring element used in drilling oil and gas wells. The bit consists of a cutting element and a circulating element. The cutting element is steel teeth, tungsten carbide buttons, industrial diamonds, or polycrystalline diamonds (PDCs). The circulating element permits the passage of drilling fluid and utilizes the hydraulic force of the fluid stream to improve drilling rates. In rotary drilling, several drill collars are joined to the bottom end of the drill pipe column, and the bit is attached to the end of the drill collars.

bleed *v*: to drain off liquid or gas, generally slowly, through a valve called a bleeder. To bleed down, or bleed off, means to release pressure slowly from a well or from pressurized equipment.

blind drilling *n*: a drilling operation in which the drilling fluid is not returned to the surface; rather, it flows into an underground formation. Sometimes blind-drilling techniques are resorted to when lost circulation occurs.

blooey line *n*: the discharge pipe from a well being drilled by air drilling. The blooey line is used to conduct the air or gas used for circulation away from the rig to reduce the fire hazard as well as to transport the cuttings a suitable distance from the well. See *air drilling*.

blowout *n*: an uncontrolled flow of gas, oil, or other well fluids into the atmosphere or into an underground formation. A blowout, or gusher, can occur when formation pressure exceeds the pressure applied to it by the column of drilling fluid. See *kick*.

blowout preventer *n*: one of several valves installed at the wellhead to prevent the escape of pressure either in the annular space between the casing and the drill pipe or in open hole (i.e., hole with no drill pipe) during drilling or completion operations. Blowout preventers on land rigs are located beneath the rig at the land's surface; on jackup or platform rigs, at the water's surface; and on floating offshore rigs, on the seafloor.

BOP *abbr*: blowout preventer.

bottomhole *n*: the lowest or deepest part of a well. *adj*: pertaining to the bottom of the wellbore.

bottomhole pressure *n*: the pressure at the bottom of a borehole. It is caused by the hydrostatic pressure of the wellbore fluid and, sometimes, by any back-pressure held at the surface, as when the well is shut in with blowout preventers. When mud is being circulated, bottomhole pressure is the hydrostatic pressure plus the remaining circulating pressure required to move the mud up the annulus.

break circulation *v*: to start the mud pump for restoring circulation of the mud column. Because the stagnant drilling fluid has thickened or gelled during the period of no circulation, high pump pressure is usually required to break circulation.

breakover *n*: the change in the chemistry of a mud from one type to another. Also called conversion.

bull gear *n*: the large circular gear in a mud pump that is driven by the prime mover and that, in turn, drives the connecting rods. Also called herringbone gear.

brine *n*: water that has a large quantity of salt, especially sodium chloride, dissolved in it; salt water.

bypass *n*: 1. a pipe connection around a valve or other control mechanism that is installed to permit passage of fluid through the line while adjustments or repairs are being made on the control. 2. a groove on a piston body that allows fluid to bypass the piston to relieve excess pressure.

bypass gate *n*: a device that allows the mud flow to be directed around the shale shaker.

bypass valve *n*: a valve that permits flow around a control valve, a piece of equipment, or a system.

C

C *abbr*: Celsius (formerly centigrade). See *Celsius scale*.

calcium contamination *n*: dissolved calcium ions in a drilling fluid in sufficient concentration to impart undesirable properties, such as flocculation, reduction in yield of bentonite, and increased fluid loss. See also *calcium hydroxide, calcium sulfate, gypsum, lime*.

calcium hydroxide *n*: the active ingredient of slaked (hydrated) lime, and the main constituent in cement (when wet). Referred to as "lime" in field terminology. Its symbol is $Ca(OH)_2$.

calcium sulfate *n*: a chemical compound of calcium, sulfur, and oxygen, $CaSO_4$. Although sometimes considered a contaminant of drilling fluids, it may at times be added to them to produce certain properties. Like calcium carbonate it forms scales in water-handling facilities, which may be hard to remove. See *anhydrite, gypsum*.

calcium-treated mud *n*: a freshwater drilling mud using calcium oxide (lime) or calcium sulfate (gypsum) to retard the hydrating qualities of shale and clay formations, thus facilitating drilling. Calcium-treated muds resist salt and anhydrite contamination but may require further treatment to prevent gelation (solidification) under the high temperatures of deep wells.

carboxymethyl cellulose (CMC) *n*: a nonfermenting cellulose product used in drilling fluids to combat calcium contamination and to lower the fluid loss of the mud.

case *n*: the outer cylinder of a concentric cylinder centrifuge. See *concentric cylinder centrifuge*.

catch samples *v*: to obtain cuttings for geological information as formations are penetrated by the bit. The samples are obtained from drilling fluid as it emerges from the wellbore or, in cable-tool drilling, from the bailer. Cuttings are carefully washed until they are free of foreign matter, dried, and labeled to indicate the depth at which they were obtained.

caustic *n*: see *caustic soda.*

caustic soda *n*: sodium hydroxide, NaOH. It is used to maintain an alkaline pH in drilling mud and in petroleum fractions. Also called caustic.

caving *n*: collapsing of the walls of the wellbore. Also called sloughing.

cavitation *n*: the formation and collapse of vapor- or gas-filled cavities that result from a sudden decrease and increase of pressure. Cavitation can cause mechanical damage to adjacent surfaces in meters, valves, pumps, and pipes at locations where flowing liquid encounters a restriction or change in direction.

Celsius scale *n*: the metric scale of temperature measurement used universally by scientists. On this scale, 0° represents the freezing point of water and 100° its boiling point at a barometric pressure of 760 mm. The Celsius scale was formerly called the centigrade scale; now, however, the term "Celsius" is preferred in the International System of Units (SI).

centigrade scale *n*: see *Celsius scale.*

centimetre (cm) *n*: a unit of length in the metric system equal to one-hundredth of a metre (10^{-2} metre).

centipoise *n*: a unit of viscosity.

centrifugal degasser *n*: a device for removing entrained gas from drilling mud in which a vacuum pump draws mud into degassing tubes inside the device. The tubes spin to create centrifugal force, which disperses the mud into thin layers on the wall of the tubes. A vacuum then pulls gas out inside an expansion tank.

centrifugal pump *n*: a pump with an impeller or rotor, an impeller shaft, and a casing, which discharges fluid by centrifugal force. An electric submersible pump is a centrifugal pump.

centrifuge *n*: a machine that uses centrifugal force to separate substances of varying densities, or specific gravities. In drilling, a centrifuge spins drilling mud at high speeds to obtain high centrifugal forces. Also called the shake-out or grind-out machine.

chemical *n*: a substance defined under HAZCOM (OSHA) as any element, chemical compound, or mixture of elements and/or compounds.

chemical barrel *n*: a container in which various chemicals are mixed prior to addition to drilling fluid.

chemical protective clothing *n*: clothing that is designed to protect against a specific chemical hazard (i.e., suits or aprons made of or coated with chemical-resistant materials like butyl rubber, neoprene, or polyvinyl chloride).

chemical treatment *n*: any of many processes in the oil industry that involve the use of a chemical to effect an operation. Some chemical treatments are acidizing, crude oil demulsification, corrosion inhibition, paraffin removal, scale removal, drilling fluid control, refinery and plant processes, cleaning and plugging operations, chemical flooding, and water purification.

chip hold-down pressure *n*: the holding of formation rock chips in place as a result of high differential pressure in the wellbore (i.e., pressure in the wellbore is greater than pressure in the formation). This effect limits the cutting action of the bit by retarding circulation of bit cuttings out of the hole.

choke *n*: a device with an orifice installed in a line to restrict the flow of fluids. Surface chokes are part of the Christmas tree on a well. Chokes are also used to control the rate of flow of the drilling mud out of the hole when the well is closed in with the blowout preventer and a kick is being circulated out of the hole.

choke manifold *n*: an arrangement of piping and special valves, called chokes. In rotary drilling, mud is circulated through a choke manifold when the blowout preventers are closed.

circulate *v*: to pass from one point throughout a system and back to the starting point. For example, drilling fluid is circulated out of the suction pit, down the drill pipe and drill collars, out the bit, up the annulus, and back to the pits while drilling proceeds.

circulating head *n*: an accessory attached to the top of the drill pipe or tubing to form a connection with the mud system to permit circulation of the drilling mud. In some cases, it is also a rotating head.

circulating pressure *n*: the pressure generated by the mud pumps and exerted on the drill stem.

circulating rate *n*: the volume flow rate of the circulating drilling fluid usually expressed in gallons or barrels per minute in the United States. Elsewhere, it is expressed in cubic metres per minute.

circulation *n*: the movement of drilling fluid out of the mud pits, down the drill stem, up the annulus, and back to the mud pits. See *normal circulation, reverse circulation.*

clay *n*: a group of hydrous aluminum silicate minerals (clay minerals).

Clean Water Act (CWA) *n*: a U.S. law that regulates the discharge of toxic and nontoxic pollutants into the surface waters of the United States. Under the jurisdiction of both the EPA and the Army Corps of Engineers, CWA's short-term goal is to make surface waters safe for recreation, fishing, and other uses. Its long-term goal is to eliminate *all* harmful discharges into surface waters.

clearance *n*: 1. the distance by which one object clears another. 2. the amount of space between two objects.

clear brine *n*: a drilling fluid made up mainly of chemical salts, such as sodium chloride, calcium chloride, or potassium chloride. Clear brine contains little or no clay or other solid material and is virtually transparent. It is often used when drilling into a producing formation because clear brine minimizes formation damage. See *formation damage.*

clear water drilling *n*: drilling operations in which plain water (usually salt water) is used as the circulating fluid.

closed system *n*: in circulation, a system in which no drilling fluid is discarded to a reserve pit. Drilling companies may use closed systems when environmental regulations do not permit any contaminants to be released. Companies are also discovering that closed systems can be economical when using expensive drilling muds.

cm *abbr*: centimetre.

cm² *abbr*: square centimetre.

cm³ *abbr*: cubic centimetre.

CMC *abbr*: carboxymethyl cellulose.

colloid *n*: 1. a substance whose particles are so fine that they will not settle out of suspension or solution and cannot be seen under an ordinary microscope. 2. the mixture of a colloid and the liquid, gaseous, or solid medium in which it is dispersed.

colloidal *adj*: pertaining to a colloid, i.e., involving particles so minute (less than 2 microns) that they are not visible through optical microscopes. Bentonite is an example of a colloidal clay.

colloidal phase *n*: see *solid phase*.

compressibility *n*: the change in volume per unit of volume of a liquid caused by a unit change in pressure at constant temperature.

compression *n*: the act or process of squeezing a given volume of gas into a smaller space.

compressor *n*: a device that raises the pressure of a compressible fluid such as air or gas. Compressors create a pressure differential to move or compress a vapor or a gas, consuming power in the process. They may be positive-displacement compressors or nonpositive-displacement compressors.

concentric cylinder centrifuge *n*: a mud centrifuge that consists of two cylinders rotating one within the other, and which, when mud is put into it, separates solid particles, such as barite or other solid particles, from the liquid mud.

condition *v*: to treat drilling mud with additives to give it certain properties. Sometimes the term applies to water used in boilers, drilling operations, and so on. To condition and circulate mud is to ensure that additives are distributed evenly throughout a system by circulating the mud while it is being conditioned. See also *mud conditioning*.

cone *n*: in mud solids control, another name for a hydrocyclone.

connecting rod *n*: 1. a forged-metal shaft that joins the piston of an engine to the crankshaft. 2. the metal shaft that is joined to the bull gear and crosshead of a mud pump.

connecting rod bearing *n*: the bearing between the rod and the crankshaft. Often called the rod bearing.

contaminant *n*: a material, usually a mud component, that becomes mixed with cement slurry during displacement and that affects it adversely.

continuous phase *n*: see *liquid phase*.

contractor *n*: see *drilling contractor*.

conversion *n*: see *breakover*.

corrosion *n*: any of a variety of complex chemical or electrochemical processes, e.g., rust, by which metal is destroyed through reaction with its environment.

corrosion-control agent *n*: a chemical added to drilling fluid to minimize corrosion to the drill stem.

crew *n*: 1. the workers on a drilling or workover rig, including the driller, the derrickhand, and the rotary helpers. 2. any group of oilfield workers.

crosshead *n*: the block in a mud pump that is guided to move in a straight line and serves as a connection between the pony rod and the connecting rod.

crosshead extension rod *n*: the rod connecting the crosshead and the piston rod in a mud pump. Also called pony rod.

cubic centimetre (cm^3) *n*: a commonly used unit of volume measurement in the metric system equal to 10^{-6} cubic metre, or 1 millilitre. The volume of a cube whose edge is 1 centimetre.

cubic foot (ft^3) *n*: the volume of a cube, all edges of which measure 1 foot. Natural gas in the United States is usually measured in cubic feet, with the most common standard cubic foot being measured at 60°F and 14.65 pounds per square inch absolute, although base conditions vary from state to state.

cubic metre (m^3) *n*: a unit of volume measurement in the metric system, replacing the previous standard unit known as the barrel, which was equivalent to 35 imperial gallons or 42 U.S. gallons. The cubic metre equals approximately 6.2898 barrels.

cuttings *n pl*: the fragments of rock dislodged by the bit and brought to the surface in the drilling mud. Washed and dried cuttings samples are analyzed by geologists to obtain information about the formations drilled.

cycle *n*: the number of strokes a piston makes from one intake stroke to another intake stroke.

cycle time *n*: the time it takes drilling mud to go down the drill stem, out of the bit, and up the annulus.

cylinder *n*: 1. a chamber in a pump from which the piston expels fluid. 2. the unit of an internal-combustion engine in which combustion and compression take place.

cylinder head *n*: the device used to seal the top of a cylinder. Also called head.

cylinder liner *n*: a removable, replaceable sleeve that fits into a cylinder.

D

dampener *n*: see *pulsation dampener*.

dampening *n*: the dissipation of energy in motion of any type, especially oscillatory motion, and the consequent reduction of decay of the motion.

decanting centrifuge *n*: a type of mud centrifuge in which a cone-shaped steel bowl rotates very fast. The action of the rotating bowl creates centrifugal force, which causes the mud solids to separate from the liquid.

deflocculation *n*: the dispersion of solids that have stuck together in drilling fluid, usually by means of chemical thinners. See *flocculation*.

defoamer *n*: any chemical that prevents or lessens frothing or foaming in another agent.

degasser *n*: the device used to remove unwanted gas from a liquid, especially from drilling fluid.

demulsifier *n*: a chemical with properties that cause the water droplets in a water-in-oil emulsion to merge and settle out of the oil, or oil droplets in an oil-in-water emulsion to coalesce, when the chemical is added to the emulsion. Also called emulsion breaker.

density *n*: the mass or weight of a substance per unit volume. For instance, the density of a drilling mud may be 10 pounds per gallon, 74.8 pounds/cubic foot, or 1,198.2 kilograms/cubic metre. Specific gravity, relative density, and API gravity are other units of density. See *mud weight*.

derrickhand *n*: the crew member who handles the upper end of the drill string as it is being hoisted out of or lowered into the hole. He or she is also responsible for the circulating machinery and the conditioning of the drilling fluid.

desander *n*: a centrifugal device for removing sand from drilling fluid to prevent abrasion of the pumps. It may be operated mechanically or by a fast-moving stream of fluid inside a special cone-shaped vessel, in which case it is sometimes called a hydrocyclone. Compare *desilter*.

desilter *n*: a centrifugal device for removing very fine particles, or silt, from drilling fluid to keep the amount of solids in the fluid at the lowest possible point. Usually, the lower the solids content of mud, the faster is the rate of penetration. The desilter works on the same principle as a desander. Compare *desander*.

diameter *n*: the distance across a circle, measured through its center. In the measurement of pipe diameters, the inside diameter is that of the interior circle and the outside diameter that of the exterior circle.

differential flowmeter *n*: an electronic instrument that measures the difference between the rate of flow of the mud going into the hole and the rate of flow of the mud out of the hole. If the rate of flow coming out of the hole increases, it may be a sign that the well has kicked.

differential pressure *n*: the difference between two fluid pressures; for example, the difference between the pressure in a reservoir and in a wellbore drilled in the reservoir, or between atmospheric pressure at sea level and at 10,000 feet (3,048 metres). Also called pressure differential.

differential sticking *n*: a condition in which the drill stem becomes stuck against the wall of the wellbore because part of the drill stem (usually the drill collars) has become embedded in the filter cake. Necessary conditions for differential-pressure sticking, or wall sticking, are a permeable formation and a pressure differential across a nearly impermeable filter cake and drill stem. Also called wall sticking. See *differential pressure*, *filter cake*.

direct-indicating viscometer *n*: a rotational device powered by means of an electric motor or handcrank. Used to determine the apparent viscosity, plastic viscosity, yield point, and gel strengths of drilling fluids. Commonly called a "V-G meter."

directional drilling *n*: 1. intentional deviation of a wellbore from the vertical. Although wellbores are normally drilled vertically, it is sometimes necessary or advantageous to drill at an angle from the vertical. Controlled directional drilling makes it possible to reach subsurface areas laterally remote from the point where the bit enters the earth. It often involves the use of deflection tools.

discharge line *n*: a line through which drilling mud travels from the mud pump to the standpipe on its way to the wellbore.

discharge valve *n*: on a mud pump, the valve that opens to allow mud to be pushed out of the pump (discharged) by the pistons moving in the liners.

dispersant *n*: a substance added to cement that chemically wets the cement particles in the slurry, allowing the slurry to flow easily without much water.

ditch *n*: a trench or channel made in the earth, usually to bury pipeline, cable, and so on. On a drilling rig, the mud flow channel from the conductor-pipe outlet is often called a ditch. See *mud return line*.

double-acting *adj*: in reference to a mud pump, the action of the piston moving mud through the cylinder on both its forward and backward stroke. Compare *single-acting*.

downhole *adj, adv*: pertaining to the wellbore.

downhole motor *n*: a drilling tool made up in the drill string directly above the bit. It causes the bit to turn while the drill string remains fixed. It is used most often as a deflection tool in directional drilling, where it is made up between the bit and a bent sub (or, sometimes, the housing of the motor itself is bent). Two principal types of downhole motor are the positive-displacement motor and the downhole turbine motor. Also called mud motor.

downstream *adv, adj*: in the direction of flow in a stream of fluid moving in a line.

downtime *n*: 1. time during which rig operations are temporarily suspended because of repairs or maintenance. 2. time during which a well is off production.

drilled solids *n pl*: the fine particles in drilling mud drilled by the bit.

driller *n*: the employee directly in charge of a drilling or workover rig and crew. The driller's main duty is operation of the drilling and hoisting equipment, but this crew member is also responsible for downhole condition of the well, operation of downhole tools, and pipe measurements.

drill-in fluid *n*: a drilling fluid specially formulated to minimize formation damage as the borehole penetrates the producing zone. See *formation damage.*

drilling contractor *n*: an individual or group of individuals who own a drilling rig or mast and contract their services for drilling wells to a certain depth.

drilling crew *n*: a driller, a derrickhand, and two or more helpers who operate a drilling or workover rig for one tour each day.

drilling fluid *n*: a liquid, air, or natural gas that is circulated through the wellbore during rotary drilling operations.

drilling mud *n*: a liquid drilling fluid containing additives to alter its properties. See *drilling fluid, mud.*

drilling rate *n*: the speed with which the bit drills the formation; usually called the rate of penetration (ROP).

drill pipe *n*: seamless steel or aluminum pipe made up in the drill stem between the kelly or top drive on the surface and the drill collars on the bottom. During drilling, it is usually rotated while drilling fluid is circulated through it.

drill stem *n*: all members in the assembly used for rotary drilling from the swivel to the bit, including the kelly, drill pipe and tool joints, drill collars, stabilizers, and various specialty items. Compare *drill string.*

drill string *n*: the column, or string, of drill pipe with attached tool joints that transmits fluid and rotational power from the kelly to the drill collars and bit. Often, especially in the oil patch, the term is loosely applied to both drill pipe and drill collars. Compare *drill stem.*

duck's nest *n*: a relatively small, excavated earthen pit into which are channeled quantities of drilling mud that overflow the usual pits.

duplex pump *n*: a reciprocating pump with two pistons or plungers. In the past the duplex, double-acting pump was the main type used as a mud pump on drilling rigs, but it has been almost entirely replaced by the triplex pump. See *double-acting, reciprocating pump.*

E

eductor *n*: a nozzlelike device that increases the velocity of a fluid. It is often used to mix two fluids or a fluid and solids.

effluent *n*: in a hydrocyclone, the liquid that exits the cone through the vortex finder. Also called overflow.

efficiency *n*: the ratio of useful energy produced by an engine to the energy put into it.

emulsifier *n*: a material that causes water and oil to form an emulsion. Water normally occurs separately from oil; if, however, an emulsifying agent is present, the water becomes dispersed in the oil as tiny droplets. Or, rarely, the oil may be dispersed in the water. In either case, the emulsion must be treated to separate the water and the oil. See also *demulsifier*.

emulsion *n*: a mixture in which one liquid, termed the dispersed phase, is uniformly distributed (usually as minute globules) in another liquid, called the continuous phase or dispersion medium. In an oil-in-water emulsion, the oil is the dispersed phase and the water the dispersion medium; in a water-in-oil emulsion, the reverse holds.

encapsulating agent *n*: a chemical in drilling mud that surrounds (encapsulates) drilled solids.

entrained *adj*: drawn in and transported by the flow of a fluid.

entrained gas *n*: formation gas that enters the drilling fluid in the annulus. See *gas-cut mud.*

environment *n*: 1. the sum of the physical, chemical, and biological factors that surround an organism. 2. the water, air, and land and the interrelationship that exists among and between water, air, and land and all living things. 3. as defined by the U.S. government, the navigable waters, the waters of the contiguous zone, the ocean waters, and any other surface water, groundwater, drinking water supply, land surface, subsurface strata, or ambient air within the United States.

Environmental Protection Agency (EPA) *n*: a U.S. agency that was created in 1970 from a variety of existing agencies. The EPA administers air pollution, water pollution, pesticide, solid waste, noise control, drinking water, and toxic substances acts. It also has major research responsibilities. Address: 401 M Street, SW; Washington, DC 20460; (202) 260-2090.

explosion-proof motor *n*: a motor with an enclosure designed to contain an internal explosion and to prevent ignition of surrounding gases or vapors by sparks that may occur in the motor.

extension rod *n*: see *crosshead extension rod.*

F

F *abbr*: Fahrenheit. See *Fahrenheit scale.*

Fahrenheit scale *n*: a temperature scale devised by Gabriel Fahrenheit, in which 32° represents the freezing point and 212° the boiling point of water at standard sea-level pressure.

Fan V-G™ meter *n*: trade name of a device used to record and measure at different speeds the flow properties of plastic fluids (such as the viscosity and gel strength of drilling fluids).

fatigue *n*: the tendency of material such as a metal to break under repeated cyclic loading at a stress considerably less than the tensile strength shown in a static test.

filter *n*: a porous medium through which a fluid is passed to separate particles of suspended solids from it.

filter cake *n*: 1. compacted solid or semisolid material remaining on a filter after pressure filtration of mud with a standard filter press. Thickness of the cake is reported in thirty-seconds of an inch or in millimetres. 2. the layer of concentrated solids from the drilling mud or cement slurry that forms on the walls of the borehole opposite permeable formations; also called wall cake or mud cake.

filter cake thickness *n*: a measurement of the solids deposited on filter paper in thirty-seconds of an inch during a standard 30-minute API filter test. In certain areas the filter cake thickness is a measurement of the solids deposited on filter paper for 7.5 minutes.

filter press *n*: a device used to test the filtration properties of drilling mud. See *filtration qualities.*

filtration *n*: the process of filtering a fluid.

filtration loss *n*: the escape of the liquid part of a drilling mud into permeable formations. Also called fluid loss.

filtration qualities *n pl*: the filtration characteristics of a drilling mud. In general, these qualities are inverse to the thickness of the filter cake deposited on the face of a porous medium and the amount of filtrate allowed to escape from the drilling fluid into or through the medium.

flange *n*: a projecting rim or edge (as on pipe fittings and openings in pumps and vessels), usually drilled with holes to allow bolting to other flanged fittings.

flare *v*: to dispose of surplus combustible vapors by burning them in the atmosphere.

flare pit *n*: on land rigs, an earthen pit dug at the end of the flare line. Gas is flared over the flare pit to protect the surrounding area from heat and fire.

flocculant *n*: see *flocculating agent.*

flocculating agent *n*: material or chemical agent that enhances flocculation. Also called flocculant.

flocculation *n*: the coagulation of solids in a drilling fluid, produced by special additives or by contaminants.

flocs *abbr, n pl*: flocculates.

flooded suction *n*: the condition of keeping enough liquid available to a pump's suction so that no air is drawn in with the liquid.

flow gate *n*: a device on the possum belly that can be moved to regulate mud flow to the shale shakers.

flow rate *n*: the speed, or velocity, of fluid flow through a pipe or vessel.

flow sensor *n*: a tool inserted into a pipeline or other container that can sense the flow of fluid within the container.

fluid *n*: a substance that flows and yields to any force tending to change its shape. Liquids and gases are fluids.

fluid cutting *n*: see *washout*.

fluid end *n*: the portion or end of a pump that contains the parts involved in moving the fluid (such as liners and pistons) as opposed to the end that produces the power for movement.

fluid-end body *n*: the steel body in a reciprocating pump that has machined cylinders, openings for the valves, and fluid passageways.

fluid knock *n*: a pressure concussion caused by suddenly stopping the flow of liquids in a closed container. Also called water hammer or hydraulic hammer.

fluid loss *n*: the unwanted migration of the liquid part of the drilling mud or cement slurry into a formation, often minimized or prevented by the blending of additives with the mud or cement.

foam *n*: a two-phase system, similar to an emulsion, in which the dispersed phase is a gas or air.

foam drilling *n*: see *mist drilling*.

foaming agent *n*: a chemical used to lighten the water column in gas wells, in oilwells producing gas, and in drilling wells in which air or gas is used as the drilling fluid so that the water can be forced out with the air or gas to prevent its impeding the production or drilling rate. See *mist drilling*.

formation *n*: a bed or deposit composed throughout of substantially the same kind of rock; often a lithologic unit. Each formation is given a name, frequently as a result of the study of the formation outcrop at the surface and sometimes based on fossils found in the formation.

formation damage *n*: the reduction of permeability in a reservoir rock caused by the invasion of drilling fluid and treating fluids to the section adjacent to the wellbore. Often called skin damage.

formation pressure *n*: the force exerted by fluids in a formation, recorded in the hole at the level of the formation with the well shut in. Also called reservoir pressure or shut-in bottomhole pressure.

fresh water *n*: 1. water that has little or no salt dissolved in it. 2. underground water, generally located near the surface, that does not contain a large amount of salt and from which most underground drinking water supplies are drawn. 3. inland surface water, such as lakes, streams, and ponds, that is not salty.

friction *n*: resistance to movement created when two surfaces are in contact. When friction is present, movement between the surfaces produces heat.

frictional resistance *n*: the opposition to flow created by a fluid when it flows through a line or other container. Frictional resistance occurs within the fluid itself and it is created by the walls of the pipe or container as the fluid flows past them.

friction loss *n*: a reduction in the pressure of a fluid caused by its motion against an enclosed surface (such as a pipe). As the fluid moves through the pipe, friction between the fluid and the pipe wall and within the fluid itself creates a pressure loss. The faster the fluid moves, the greater are the losses.

ft *abbr*: foot.

ft/min *abbr*: feet per minute.

ft^2 *abbr*: square foot.

ft^3 *abbr*: cubic foot.

ft^3/min *abbr*: cubic feet per minute.

funnel viscosity *n*: viscosity as measured by the Marsh funnel, based on the number of seconds it takes for 61 cubic inches (1,000 cubic centimetres) of drilling fluid to flow through the funnel.

G

gal *abbr*: gallon.

gallon *n*: a unit of measure of liquid capacity that equals 3.785 litres and has a volume of 231 cubic inches (0.00379 cubic metres). A gallon of water weighs 8.34 pounds (3.8 kilograms) at 60°F (16°C). The imperial gallon, formerly used in Great Britain, equals approximately 1.2 U.S. gallons.

gal/min *abbr*: gallons per minute.

gas *n*: a compressible fluid that completely fills any container in which it is confined.

gas buster *n*: a mud-gas separator.

gas-cut mud *n*: a drilling mud that contains entrained formation gas, giving the mud a characteristically fluffy texture. When entrained gas is not released before the fluid returns to the well, the weight or density of the fluid column is reduced. Because a large amount of gas in mud lowers its density, gas-cut mud must be treated to reduce the chance of a kick.

gas cutting *n*: a process in which gas becomes entrained in a liquid.

gas drilling *n*: a method of drilling that uses natural gas as the drilling fluid. See also *air* drilling.

gel *n*: 1. a semisolid, jellylike state assumed by some colloidal dispersions at rest. When agitated, the gel converts to a fluid state. 2. a nickname for bentonite. *v*: to take the form of a gel; to set.

gel strength *n*: a measure of the ability of a colloidal dispersion to develop and retain a gel form, based on its resistance to shear. The gel, or shear, strength of a drilling mud determines its ability to hold solids in suspension. Sometimes bentonite and other colloidal clays are added to drilling fluid to increase its gel strength.

gland *n*: a device used to form a seal around a reciprocating or rotating rod (as in a pump) to prevent fluid leakage. Specifically, the movable part of a stuffing box by which the packing is compressed. See *stuffing box*.

gland packing *n*: material that creates a seal around a reciprocating or rotating rod.

gland-packing nut *n*: a threaded device the sides of which are arranged so that a wrench can be fitted onto them and used to retain the gland packing in place around a rod. See *gland packing*.

gooseneck *n*: the curved connection between the rotary hose and the swivel. See *swivel.*

gpm *abbr*: gallons per minute, when referring to rate of flow.

groundwater *n*: water that seeps through soil and fills pores of underground rock formations; the source of water in springs and wells.

gumbo *n*: any relatively sticky formation (such as clay) encountered in drilling.

gyp *n*: (slang) gypsum.

gyp mud *n*: drilling mud that is treated with gypsum to provide a source of soluble calcium in the filtrate to obtain desirable mud properties for drilling in shale or clay formations.

gypsum *n*: a naturally occurring crystalline form of calcium sulfate in which each molecule of calcium sulfate is combined with two molecules of water. See *anhydrite, calcium sulfate.*

H

Hazard Communication Standard (HAZCOM, HCS, or the "Employee Right to Know") *n*: an OSHA standard that guarantees employees the right to know about chemical hazards on the job and how to protect themselves from those hazards. Under HAZCOM, all manufacturers and employers must prepare a written hazard communication program; prepare a list of all hazardous materials in the workplace; label all containers of hazardous materials in the workplace; collect and maintain a material safety data sheet (MSDS) for each hazardous material present in the workplace; and provide employee training on specific topics related to hazardous substances.

hazardous *adj*: involving or exposing one to risk. The lists of material or waste that are considered hazardous vary from agency to agency and from regulation to regulation: hazardous materials in transport are regulated by DOT; hazardous substances in the workplace are regulated by OSHA; hazardous waste is regulated under EPA's RCRA; toxic substances are regulated under EPA's TSCA; hazardous air pollutants are regulated under EPA's CAA; and so on. The hazardous list for that regulation is tailored to that purpose.

hazardous chemical *n*: 1. (OSHA) (HAZCOM) any chemical that is a physical hazard or a health hazard. 2. (SARA) any hazardous chemical as defined under 29 CFR 1910.1200(c), except those that are regulated by other agencies or laws.

hazardous materials (HAZMAT) *n pl*: (DOT) substances or materials in quantities or forms that may pose an unreasonable risk to health, safety, or property when stored, transported, or used in commerce.

hazardous substance *n*: 1. any substance designated under CWA or CERCLA as posing a threat to waterways and the environment when released. 2. (CERCLA) any substance designated in 40 CFR 302.

hazardous waste *n*: 1. any solid waste ("solid" includes any solid, liquid, semisolid, or contained gaseous material) resulting from industrial, commercial, mining, or agricultural operations, or from community activities, that meets certain characteristics of hazard (i.e., ignitable, corrosive, reactive, toxic) or that is listed as a waste from specific or nonspecific sources or that is a listed commercial chemical product or manufacturing intermediate that is sometimes discarded. 2. (RCRA) discarded materials regulated by the EPA because of public health and safety concerns. 3. (HAZWOPER) a waste or combination of wastes as defined in 40 CFR 261.3, or those substances defined as hazardous wastes in 49 CFR 171.8.3. 4. (CERCLA) those wastes listed in 40 CFR 261.3.

Hazardous Waste Operations and Emergency Response Standard (HAZWOPER) *n*: an OSHA standard that is concerned primarily with worker safety in emergency response situations. HAZWOPER requires employers to protect the safety and health of three specific groups of workers: those involved in emergency response or cleanup at hazardous waste sites; those involved in emergency response at treatment, storage, and disposal (TSD) sites; and those involved in emergency response to incidents involving hazardous substances.

hazard warning *n*: (OSHA) any words, pictures, symbols, or combination thereof appearing on a label or other appropriate form of warning that convey the hazard(s) of the chemical(s) in the container(s).

HAZCOM *abbr*: Hazard Communication Standard.

HAZMAT *abbr*: hazardous material.

head *n*: 1. the height of a column of liquid required to produce a specific pressure. See *hydraulic head*. 2. for centrifugal pumps, the velocity of flowing fluid converted into pressure expressed in feet or metres of flowing fluid. Also called velocity head. 3. that part of a machine (such as a pump or an engine) that is on the end of the cylinder opposite the crankshaft. Also called cylinder head.

herringbone gear *n*: see *bull gear*.

high-pH mud *n*: a drilling fluid with a pH range above 10.5, i.e., a high-alkalinity mud.

hold-down pressure *n*: hydrostatic pressure developed by the weight of the drilling fluid exerted on the bottom of the hole that tends to prevent cuttings from moving up the annulus.

horizontal drilling *n*: deviation of the borehole at least 80° from vertical so that the borehole penetrates a productive formation in a manner parallel to the formation. A single horizontal hole can effectively drain a reservoir and eliminate the need for several vertical boreholes.

horsepower *n*: a unit of measure of work done by a machine. One horsepower equals 33,000 foot-pounds per minute. (Kilowatts are used to measure power in the international, or SI, system of measurement.)

hydraulic *adj*: 1. of or relating to water or other liquid in motion. 2. operated, moved, or effected by water or liquid.

hydraulic hammer *n*: see *fluid knock*.

hydraulic head *n*: the force exerted by a column of liquid expressed by the height of the liquid above the point at which the pressure is measured. Although "head" refers to distance or height, it is used to express pressure, since the force of the liquid column is directly proportional to its height. Also called head or hydrostatic head. Compare *hydrostatic pressure*.

hydraulics *n*: 1. the branch of science that deals with practical applications of water or other liquid in motion. 2. the planning and operation of a rig hydraulics program, coordinating the power of circulating fluid at the bit with other aspects of the drilling program so that bottomhole cleaning is maximized.

hydrocyclone *n*: a cone-shaped separator for separating various sizes of particles and liquid by centrifugal force. See *desander*, *desilter*.

hydrophilic *adj*: tending to absorb water.

hydrostatic pressure *n*: the force exerted by a body of fluid at rest. It increases directly with the density and the depth of the fluid and is expressed in pounds per square inch or kilopascals. The hydrostatic pressure of fresh water is 0.433 pounds per square inch per foot (9.792 kilopascals per metre) of depth. In drilling, the term refers to the pressure exerted by the drilling fluid in the wellbore. In a water drive field, the term refers to the pressure that may furnish the primary energy for production.

I

impeller *n*: a set of mounted blades used to impart motion to a fluid (e.g., the rotor of a centrifugal pump).

instrumentation *n*: a device or assembly of devices designed for one or more of the following functions: to measure operating variables (such as pressure, temperature, rate of flow, and speed of rotation); to indicate these phenomena with visible or audible signals; to record them; to control them within a predetermined range; and to stop operations if the control fails. Simple instrumentation might consist of an indicating pressure gauge only. In a completely automatic system, desired ranges of pressure, temperature, and so on are predetermined and preset.

intake valve *n*: 1. the cam-operated mechanism on an engine through which air and sometimes fuel are admitted to the cylinder. 2. on a mud pump, the valve that opens to allow mud to be drawn into the pump by the pistons moving in the liners.

invert-emulsion mud *n*: an oil mud in which fresh or salt water is the dispersed phase and diesel, crude, or some other oil is the continuous phase. Also called invert-oil mud. See *oil mud*.

J

jet hopper *n*: a large funnel- or cone-shaped device into which dry components (such as powdered clay or cement) can be poured to mix uniformly with water or other liquids. The liquid is injected through a nozzle at the bottom of the hopper. The resulting mixture may be drilling mud to be used as the circulating fluid in a rotary drilling operation, or it may be cement slurry to be used in bonding casing to the borehole.

jet siphon *n*: a special pipe installed near the bottom of a mud tank that turns upward from the tank's bottom and is inserted into the end of a larger pipe. When mud is pumped through the small pipe, the pressure lifts mud out of the tank through the larger pipe. This lifting effect is called a siphon effect.

jetting *n*: the process of periodically removing a portion or all of the water, mud, and/or solids from the tanks, usually by means of pumping through a jet nozzle arrangement.

K

kelly *n*: the heavy steel tubular device, usually four- or six-sided suspended from the swivel through the rotary table and connected to the top joint of drill pipe to turn the drill stem as the rotary table turns. It has a bored passageway that permits fluid to be circulated into the drill stem and up the annulus, or vice versa.

kelly hose *n*: see *rotary hose*.

kick *n*: an entry of water, gas, oil, or other formation fluid into the wellbore during drilling. It occurs because the pressure exerted by the column of drilling fluid is not great enough to overcome the pressure exerted by the fluids in the formation drilled. If prompt action is not taken to control the kick, or kill the well, a blowout may occur.

kill *v*: 1. in drilling, to control a kick by taking suitable preventive measures (e.g., to shut in the well with the blowout preventers, circulate the kick out, and increase the weight of the drilling mud). 2. in production, to stop a well from producing oil and gas so that reconditioning of the well can proceed. Production is stopped by circulating a kill fluid into the hole.

kill fluid *n*: drilling mud of a weight great enough to equal or exceed the pressure exerted by formation fluids.

kill line *n*: a pipe attached to the blowout preventer stack, into which mud or cement can be pumped to overcome the pressure of a kick. Sometimes used when normal kill procedures (circulating kill fluids down the drill stem) are not sufficient.

kill mud *n*: see *kill fluid.*

kill rate *n*: the speed, or velocity, of the mud pump used when killing a well. Usually measured in strokes per minute, it is considerably slower than the rate used for normal operations.

kill rate pressure *n*: the pressure exerted by the mud pump (and read on the standpipe or drill pipe pressure gauge) when the pump's speed is reduced to a speed lower than that used during normal drilling. A kill rate pressure or several kill rate pressures are established for use when a kick is being circulated out of the wellbore. Also called p-low.

kill sheet *n*: a printed form that contains blank spaces for recording information about killing a well. It is provided to remind personnel of the necessary steps to take to kill a well.

kilograms per cubic centimetre (kg/m^3) *n*: a measure of the density, or weight, of a fluid (such as drilling mud).

kilopascal (kPa) *n*: 1,000 pascals. The SI metric unit of measurement for pressure and stress and a component in the measurement of viscosity. A pascal is equal to a force of 1 newton acting on an area of 1 square metre.

kPa *abbr*: kilopascal.

L

lantern ring *n*: the middle ring in a liner packing in a reciprocating pump. A notch on the lantern ring lines up with the grease fitting so that when lubricant is injected, the ring distributes it between the gland packing rings.

lignins *n pl*: naturally occurring special lignites, e.g., leonardite, that are produced by strip mining from special lignite deposits. Used primarily as thinners and emulsifiers.

lignosulfonate *n*: an organic drilling fluid additive derived from by-products of a papermaking process using sulfite. It minimizes fluid loss and reduces mud viscosity.

lime *n*: a caustic solid that consists primarily of calcium oxide (CaO). Many forms of CaO are called lime, including the various chemical and physical forms of quicklime, hydrated lime, and even calcium carbonate. Limestone is sometimes called lime.

lime mud *n*: a drilling mud that is treated with lime to provide a source of soluble calcium in the filtrate to obtain desirable mud properties for drilling in shale or clay formations.

liner *n*: see *cylinder liner*.

liquid *n*: a state of matter in which the shape of the given mass depends on the containing vessel, but the volume of the mass is independent of the vessel. A liquid is a fluid that is almost incompressible. Also called continuous phase.

liquid phase *n*: in drilling fluids, that part of the fluid that is liquid. Normally, the liquid phase of a drilling fluid is water, oil, or a combination of water and oil.

lost circulation *n*: the quantities of whole mud lost to a formation, usually in cavernous, fissured, or coarsely permeable beds. Evidenced by the complete or partial failure of the mud to return to the surface as it is being circulated in the hole. Lost circulation can lead to a blowout and, in general, can reduce the efficiency of the drilling operation. Also called lost returns.

lost circulation additives *n pl*: materials added to the mud in varying amounts to control or prevent lost circulation. Classified as fiber, flake, or granular.

lost circulation material (LCM) *n*: a substance added to cement slurries or drilling mud to prevent the loss of cement or mud to the formation. See *lost circulation additives*.

low clay-solids mud *n*: heavily weighted muds whose high solids content (a result of the large amounts of barite added) necessitates the reduction of clay solids.

low-solids mud *n*: a drilling mud that contains a minimum amount of solid material and that is used in rotary drilling when possible because it can provide fast drilling rates.

lubricant *n*: a substance—usually petroleum-based—that is used to reduce friction between two moving parts.

M

m *abbr*: metre.

m² *abbr*: square metre.

m³ *abbr*: cubic metre.

manifold *n*: 1. an accessory system of piping to a main piping system (or another conductor) that serves to divide a flow into several parts, to combine several flows into one, or to reroute a flow to any one of several possible destinations. 2. a pipe fitting with several side outlets to connect it with other pipes.

manifold capacity *n*: the number of cones needed in a mud cleaner to process the full volume of circulating mud.

Marsh funnel *n*: a calibrated funnel used in field tests to determine the viscosity of drilling mud.

material safety data sheet (MSDS) *n*: a reference document prepared by the chemical manufacturer and required under HAZCOM to be kept with the material during shipment and distributed in the workplace for any employee who may work with that material. The information on the MSDS may take a variety of forms but it must include the identity of the chemical, chemicals, or mixtures as listed on the container label; the physical and chemical characteristics of the material; the physical hazards of the material; the health hazards of the material; the primary routes of entry into the human body; the permissible exposure limits, if available; whether the material has been found to be a carcinogen, a potential carcinogen, or is regulated by OSHA; any generally applicable precautions for safe handling; any generally applicable control measures; emergency first aid procedures; the date the MSDS was prepared; and the name, address, and telephone number of the material source or another responsible party who can provide additional information and emergency procedures for the material if necessary.

mesh *n*: a measure of the openings of a woven material, screen, or sieve; e.g., a 200-mesh sieve has 200 openings per linear inch. A 200-mesh screen with a wire diameter of 0.0021 inch (0.0533 millimetre) has an opening of 0.074 millimetre (74 microns), or will pass a particle of 74 microns. See *micron*.

micron *n*: a unit of length equal to one millionth part of a metre, or one thousandth part of a millimetre, or about 0.000039 inch. Also called a micrometre.

millimetre (mm) *n*: a measurement unit in the metric system equal to 10^{-3} metre (0.001 metre). It is used to measure pipe and bit diameter, nozzle size, liner length and diameter, and cake thickness.

min *abbr*: minute.

mist drilling *n*: a drilling technique that uses air or gas to which a foaming agent has been added. Also called foam drilling.

mixing mud *n*: preparation of drilling fluids from a mixture of water and other fluids and one or more of the various dry mud-making materials such as clay and chemicals.

mixing tank *n*: any tank or vessel used to mix components of a substance (as in the mixing of additives with drilling mud).

mix mud *v*: to prepare drilling fluids from a mixture of water or other liquids and any one or more of the various dry mud-making materials (such as clay, weighting materials, and chemicals).

mm *abbr*: millimetre.

mm^2 *abbr*: square millimetre.

mm^3 *abbr*: cubic millimetre.

montmorillonite *n*: a clay mineral often used as an additive to drilling mud. It is a hydrous aluminum silicate capable of reacting with such substances as magnesium and calcium. See also *bentonite*.

mud *n*: the liquid circulated through the wellbore during rotary drilling and workover operations. Although it was originally a suspension of earth solids (especially clays) in water, the mud used in modern drilling operations is a more complex, three-phase mixture of liquids, reactive solids, and nonreactive solids. The liquid phase may be fresh water, diesel oil, or crude oil and may contain one or more conditioners. See *drilling fluid*, *drilling mud*.

mud additive *n*: any material added to drilling fluid to change some of its characteristics or properties.

mud agitator *n*: see *agitator*.

mud analysis *n*: examination and testing of drilling mud to determine its physical and chemical properties.

mud balance *n*: a beam balance consisting of a cup and a graduated arm carrying a sliding weight and resting on a fulcrum. It is used to determine the density or weight of drilling mud.

mud cake *n*: see *filter cake*, *wall cake*.

mud circulation *n*: the process of pumping mud downward to the bit and back up to the surface in a drilling or workover operation. See *normal circulation*, *reverse circulation*.

mud cleaner *n*: a cone-shaped device, a hydrocyclone, designed to remove very fine solid particles from the drilling mud.

mud column *n*: the borehole when it is filled or partially filled with drilling mud.

mud conditioning *n*: the treatment and control of drilling mud to ensure that it has the correct properties. Conditioning may include the use of additives, the removal of sand or other solids, the removal of gas, the addition of water, and other measures to prepare the mud for conditions encountered in a specific well.

mud density *n*: see *mud weight*.

mud density recorder *n*: a device that automatically records the weight or density of drilling fluid as it is being circulated in a well.

mud engineer *n*: an employee of a drilling fluid supply company whose duty it is to test and maintain the drilling mud properties that are specified by the operator.

mud-flow indicator *n*: a device that continually measures and may record the flow rate of mud returning from the annulus and flowing out of the mud return line. If the mud does not flow at a fairly constant rate, a kick or lost circulation may have occurred.

mud-gas separator *n*: a device that removes entrained gas from drilling mud. Also called gas buster.

mud gun *n*: a device that shoots a jet of drilling mud under high pressure into the mud pit to mix additives with the mud or to agitate the mud. Rarely used now, it has been replaced by agitators.

mud hose *n*: also called kelly hose or rotary hose. See *rotary hose*.

mud hound *n*: see *mud engineer*.

mud house *n*: structure at the rig to store and shelter sacked materials used in drilling fluids.

mud inhibitor *n*: a substance, such as salt, potassium chloride, or calcium sulfate, added to drilling mud to minimize the hydration (swelling) of formations with which the mud is in contact.

mud line *n*: 1. in offshore operations, the seafloor. 2. a mud return line.

mud log *n*: a record of information derived from examination of drilling fluid and drill bit cuttings. See *mud logging*.

mud logging *n*: the recording of information derived from examination and analysis of formation cuttings made by the bit and of mud circulated out of the hole. A portion of the mud is diverted through a gas-detecting device. Cuttings brought up by the mud are examined under ultraviolet light to detect the presence of oil or gas. Mud logging is often carried out in a portable laboratory set up at the well.

mud man *n*: see *mud engineer*.

mud manifold *n*: see *pump manifold.*

mud-mixing devices *n pl*: any of several devices used to agitate, or mix, the liquids and solids that make up drilling fluid. These devices include jet hoppers, paddles, stirrers, and chemical barrels.

mud pit *n*: originally, an open pit dug in the ground to hold drilling fluid or waste materials discarded after the treatment of drilling mud. Now it is another name for a mud tank; see *mud tank*.

mud program *n*: a plan or procedure, with respect to depth, for the type and properties of drilling fluid to be used in drilling a well. Some factors that influence the mud program are the casing program and such formation characteristics as type, competence, solubility, temperature, and pressure.

mud pump *n*: a large, high-pressure reciprocating pump used to circulate the mud on a drilling rig. A typical mud pump is a triplex or a duplex pump whose pistons travel in replaceable liners and are driven by a crankshaft actuated by an engine or a motor. Also called a slush pump.

mud report *n*: a special form that is filled out by the mud engineer to record the properties of the drilling mud used while a well is being drilled.

mud return line *n*: a trough or pipe that is placed between the surface connections at the wellbore and the shale shaker and through which drilling mud flows on its return to the surface from the hole. Also called flow line.

mud screen *n*: see *shale shaker*.

mud solids *n pl*: the solid components of drilling mud. They may be added intentionally (e.g., barite), or they may be introduced into the mud from the formation as the bit drills ahead. The term is usually used to refer to the latter.

mud still *n*: instrument used to distill oil, water, and other volatile materials in a mud to determine oil, water, and total solids contents in volume-percent. Also called retort.

mud tank *n*: a steel tank to hold drilling fluid or waste materials discarded after the treatment of drilling mud. For some drilling operations, mud pits are used for suction to the mud pumps, settling of mud sediments, and storage of reserve mud. They may still be referred to as mud pits. Also called slush tank.

mud volume totalizer *n*: see *pit-level indicator.*

mud weight *n*: a measure of the density of a drilling fluid expressed as pounds per gallon, pounds per cubic foot, or kilograms per cubic metre. Mud weight is directly related to the amount of pressure the column of drilling mud exerts at the bottom of the hole.

mud weight indicator *n*: an instrument that automatically determines a mud's weight (density) and displays it on a readout for rig personnel.

N

natural clays *n pl*: clays that are encountered when drilling various formations; they may or may not be incorporated purposely into the mud system.

natural gas *n*: a highly compressible, highly expansible mixture of hydrocarbons with a low specific gravity and occurring naturally in a gaseous form. Besides hydrocarbon gases, natural gas may contain appreciable quantities of nitrogen, helium, carbon dioxide, hydrogen sulfide, and water vapor. Although gaseous at normal temperatures and pressures, the gases making up the mixture that is natural gas are variable in form and may be found either as gases or as liquids under suitable conditions of temperature and pressure.

natural mud *n*: a drilling fluid containing essentially clay and water; no special or expensive chemicals or conditioners are added. Also called conventional mud.

nonreactive phase *n*: that part of a liquid drilling mud that consists of solids or other chemicals that do not react with the liquid (or other chemicals in the liquid) part of the mud. Barite, for example, is nonreactive.

normal circulation *n*: the smooth, uninterrupted circulation of drilling fluid down the drill stem, out the bit, up the annular space between the pipe and the hole, and back to the surface. Compare *reverse circulation*.

normal formation pressure *n*: formation fluid pressure equivalent to about 0.465 pounds per square inch per foot (10.5 kilopascals per metre) of depth from the surface. If the formation pressure is 4,650 pounds per square inch (32,062 kilopascals) at 10,000 feet (3,048 metres), it is considered normal.

nozzle *n*: a passageway through jet bits that causes the drilling fluid to be ejected from the bit at high velocity. The jets of mud clear the bottom of the hole. Nozzles come in different sizes that can be interchanged on the bit to adjust the velocity with which the mud exits the bit.

O

Occupational Safety and Health Administration (OSHA) *n*: a U.S. government agency that conducts research into the causes of occupational diseases and accidents. It is responsible for administration of the certification of respiratory safety equipment. Address: Department of Labor; 200 Constitution Avenue, NW; Washington, DC 20210; (202) 219-8148.

oil-base mud *n*: a drilling or workover fluid in which oil is the continuous phase and which contains from less than 2 percent and up to 5 percent water. This water is spread out, or dispersed, in the oil as small droplets. See *invert-emulsion mud, oil mud*.

oil-emulsion mud *n*: a water-base mud in which water is the continuous phase and oil is the dispersed phase. The oil is spread out, or dispersed, in the water in small droplets, which are tightly emulsified so that they do not settle out. Because of its lubricating abilities, an oil-emulsion mud increases the drilling rate and ensures better hole conditions than other muds. Compare *oil mud*.

oil mud *n*: a drilling mud, e.g., oil-base mud and invert-emulsion mud, in which oil is the continuous phase. It is useful in drilling certain formations that may be difficult or costly to drill with water-base mud. Compare *oil-emulsion mud*.

operator *n*: the person or company, either proprietor or lessee, actually operating an oilwell or lease, generally the oil company that engages the drilling contractor.

organophilic clay *n*: a clay treated so that it is not hydrophilic for use in oil muds. Also called organic clay, amine clay.

overbalanced drilling *n*: drilling in which the hydrostatic pressure of the mud column exceeds formation pressure.

overflow *n*: the effluent of a cone-shaped centrifuge that passes up the inside of the cone and leaves through the vortex finder. Also called effluent.

P

packing *n*: a material used in a cylinder, in the stuffing box of a valve, or between flange joints to maintain a leakproof seal.

packing gland *n*: see *gland.*

packing ring *n*: piston ring.

personal protective equipment (PPE) *n*: items to protect rig workers from hazardous chemicals, such as rubber gloves, aprons, face shields, and respirators.

pH *abbr*: an indicator of the acidity or alkalinity of a substance or solution, represented on a scale of 0–14, 0–6.9 being acidic, 7 being neither acidic nor basic (i.e., neutral), and 7.1–14 being basic. These values are based on hydrogen ion content and activity.

phase *n*: a portion of a physical system that is liquid, gas, or solid, that is homogeneous throughout, that has definite boundaries, and that can be separated from other phases. The three phases of drilling mud are the liquid or continuous phase, the reactive or colloidal phase, and the nonreactive phase.

pH control agent *n*: a chemical added to the drilling fluid to control or increase the pH of the mud. Normally, the mud should have a pH higher than seven so that it is alkaline.

pill tank *n*: a small tank on some rigs that may contain an especially viscous quantity of mud, or a pill. The driller can pump the pill of viscous mud down the hole and place it at or near a formation that is causing lost circulation problems.

pinion *n*: 1. a gear with a small number of teeth designed to mesh with a larger wheel or rack. 2. the smaller of a pair or the smallest of a train of gear wheels.

pinion gear *n*: on a mud pump's power end, the gear that, when turned by a shaft, engages the bull gear to turn it. When the bull gear turns, the pump operates.

pinion shaft *n*: a shaft inside a mud pump, which turns a small gear, driven by power bands.

piston *n*: a cylindrical sliding piece that is moved by or that moves against fluid pressure within a confining cylindrical vessel.

piston body *n*: the metal body of a piston over which the piston rubber fits.

piston pin *n*: a pin that forms a flexible link between the piston and the connecting rod. This bearing area has the highest load per square inch (square millimetre) of any in an engine, perhaps as high as 50,000 pounds per square inch (345 megapascals). Also called a wrist pin.

piston ring *n*: a yielding ring, usually metal, that surrounds a piston and maintains a tight fit inside a cylinder.

piston rod *n*: 1. a metal shaft that joins the piston to the crankshaft in an engine. 2. a metal shaft in a mud pump, one end of which is connected to the piston and the other to the pony rod.

piston rubber *n*: the flexible rubber or synthetic sealing element of a piston.

pit *n*: 1. a temporary containment, usually excavated earth, for wellbore fluids. 2. a mud tank. 3. a reserve pit.

pit level *n*: height of drilling mud in the mud tanks, or pits.

pit-level indicator *n*: one of a series of devices that continuously monitor the level of the drilling mud in the mud tanks. The indicator usually consists of float devices in the mud tanks that sense the mud level and transmit data to a recording and alarm device (a pit-volume recorder) mounted near the driller's position on the rig floor. If the mud level drops too low or rises too high, the alarm sounds to warn the driller of lost circulation or a kick.

pit watcher *n*: in offshore drilling rigs, the derrickhand or an assistant who monitors the mud tanks and drilling mud properties, mixes mud, and transfers mud from one tank to another.

plastic viscosity *n*: an absolute flow property indicating the flow resistance of certain types of fluids. It is a measure of shear stress.

plunger *n*: the rod that serves as a piston in a reciprocating pump.

polymer *n*: a substance that consists of large molecules formed from smaller molecules in repeating structural units (monomers). In oilfield operations, various types of polymers are used to thicken drilling mud, fracturing fluid, acid, water, and other liquids. See *polymer mud*.

polymer mud *n*: a drilling mud to which a polymer has been added to increase the viscosity of the mud.

pony rod *n*: the rod joined to the connecting rod and piston rod in a mud pump. Also called crosshead extension rod.

poor boy degasser *n*: usually, a mud-gas separator that is fabricated by the personnel on a drilling rig's location, or by welders employed by the drilling contractor in the contractor's storage yard. It is a steel, airtight cylinder into which drilling mud is piped to provide a space for gas in the mud to escape.

positive-displacement pump *n*: a reciprocating or a rotary pump that moves a measured quantity of liquid with each stroke of a piston or each revolution of vanes or gears.

possum belly *n*: a receiving tank situated at the end of the mud return line. The flow of mud comes into the bottom of the device and travels over baffles to control mud flow over the shale shaker.

pounds per cubic foot *n*: a measure of the density of a substance (such as drilling fluid).

pounds per gallon (ppg) *n*: a measure of the density, or weight, of a fluid (such as drilling mud).

pounds per 100 square feet *n*: a measure of force, for example, of the gel strength of drilling mud.

pounds per square inch (psi) *n*: a measure of pressure.

power band *n*: on rigs that use a compound, large V-belts between the compound and the pump's power end that drive a pinion shaft inside the pump.

power end *n*: the portion or end of a mud pump that contains the parts involved in producing the mechanical force that moves the pistons in liners to move liquid. The power end is opposite the fluid end, which has the parts that move the liquid mud.

ppg *abbr*: pounds per gallon.

pressure *n*: the force that a fluid (liquid or gas) exerts uniformly in all directions within a vessel, pipe, hole in the ground, and so forth, such as that exerted against the inner wall of a tank or that exerted on the bottom of the wellbore by a fluid. Pressure is expressed in terms of force exerted per unit of area, as pounds per square inch, or in kilopascals.

pressure gauge *n*: an instrument that measures fluid pressure and usually registers the difference between atmospheric pressure and the pressure of the fluid by indicating the effect of such pressures on a measuring element (e.g., a column of liquid, pressure in a Bourdon tube, a weighted piston, or a diaphragm).

pressure loss *n*: the drilling fluid's loss of hydraulic pressure after it leaves the pump. Some pressure is lost due to friction, but the main loss occurs when the fluid leaves the bit nozzles. See also *friction loss*.

pressure-relief valve *n*: a valve that opens at a preset pressure to relieve excessive pressures within a vessel or line. Also called a pop valve, relief valve, safety valve, or safety relief valve.

pressure surge *n*: a sudden and usually short-duration increase in pressure. When pipe or casing is run into a hole too rapidly, an increase in the hydrostatic pressure results, which may be great enough to create lost circulation.

prime *v*: to fill a pump with drilling fluid before starting it up in order to prevent air from entering the system.

producing zone *n*: the zone or formation from which oil or gas is produced.

psi *abbr*: pounds per square inch.

P-tank *n*: a tank or bin for storing dry mud additives delivered in bulk form by truck or boat.

pulsation dampener *n*: 1. any gas- or liquid-charged, chambered device that minimizes periodic increases and decreases in pressure (as from a mud pump). 2. a device used to reduce pressure pulsations in a flowing stream. Also called surge dampener.

pump *n*: a device that increases the pressure on a fluid. Types of pumps used in the circulating system include the mud pump (a reciprocating pump), centrifugal pumps, and the downhole pump.

pumping cycle *n*: see *cycle*.

pump liner *n*: a cylindrical, accurately machined, metallic section that forms the working barrel of some reciprocating pumps. Liners are an inexpensive means of replacing worn cylinder surfaces, and in some pumps they provide a method of conveniently changing the displacement and capacity of the pumps.

pump manifold *n*: an arrangement of valves and piping that permits a wide choice in the routing of suction and discharge fluids among two or more pumps. Also called mud manifold.

pump pressure *n*: fluid pressure arising from the action of a pump.

pump speed *n*: the speed, or velocity, at which a pump is run. In drilling, the pump speed is usually measured in strokes per minute.

pump stroke counter *n*: see *pump stroke indicator*.

pump stroke indicator *n*: an instrument that measures pump speed by counting the number of strokes per minute. Also called pump stroke counter.

R

rate of penetration (ROP) *n*: the speed with which the bit drills the formation.

reactive phase *n*: see *solid phase*.

reactive solids content *n*: the amount of water-absorbent material in the drilling fluid.

reciprocating pump *n*: a pump consisting of a piston that moves back and forth or up and down in a cylinder. The cylinder is equipped with inlet (suction) and outlet (discharge) valves. On the intake stroke, the suction valves are opened, and fluid is drawn into the cylinder. On the discharge stroke, the suction valves close, the discharge valves open, and fluid is forced out of the cylinder.

reciprocation *n*: a back-and-forth or up-and-down movement (as the movement of a piston in an engine or pump).

recording instrument *n*: a measuring instrument that records the value of the measured variable by marking or printing on a removable paper chart, tape, or other suitable recording material.

red-lime mud *n*: a water-base clay mud containing caustic soda and tannates to which lime has been added. Also called red mud.

red mud *n*: see *red-lime mud*.

representative sample *n*: a small portion extracted from the total volume of material that contains the same proportions of the various flowing constituents as the total volume of liquid being transferred. The precision of extraction must be equal to or better than the method used to analyze the sample.

reserve pit *n*: 1. a waste pit, usually an excavated earthen-walled pit. It may be lined with plastic or other material to prevent soil contamination. 2. (obsolete) a mud pit in which a supply of drilling fluid is stored.

reserve tank *n*: a special mud tank that holds mud that is not being actively circulated. A reserve tank usually contains a different type of mud from that which the pump is currently circulating. For example, it may store heavy mud for emergency well-control operations.

retort *n*: see *mud still*.

returns *n pl*: the mud, cuttings, and so forth, that circulate up the hole to the surface.

reverse circulation *n*: the course of drilling fluid downward through the annulus and upward through the drill stem, in contrast to normal circulation in which the course is downward through the drill stem and upward through the annulus. Seldom used in open hole, but frequently used in workover operations. Also referred to as "circulating the short way," since returns from bottom can be obtained more quickly than in normal circulation.

rig up *v*: to prepare the drilling rig for making hole, that is, to install tools and machinery before drilling is started.

ring grooves *n pl*: the grooves that hold the piston rings in pistons.

riser *n*: a pipe through which liquid travels upward.

rod packing *n*: a special material installed around a pump's piston rod that allows the rod to move back and forth yet maintains a seal to prevent mud from getting onto the rod and scoring it.

ROP *abbr*: rate of penetration.

rotary hose *n*: a reinforced flexible tube on a rotary drilling rig that conducts the drilling fluid from the standpipe to the swivel and kelly. Also called the mud hose or the kelly hose.

rotating head *n*: a sealing device used to close off the annular space around the kelly in drilling with pressure at the surface, usually installed above the main blowout preventers. A rotating head makes it possible to drill ahead even when there is pressure in the annulus that the weight of the drilling fluid is not overcoming; the head prevents the well from blowing out. It is used mainly in the drilling of for-mations that have low permeability. The rate of penetration through such formations is usually rapid.

rotating impeller *n*: in a centrifugal pump, a vanelike device that turns (rotates) to create an area of low pressure in the center of the impeller.

rotor *n*: 1. a device with vanelike blades attached to a shaft. The device turns or rotates when the vanes are struck by a fluid directed there by a stator. 2. the rotating part of an induction-type alternating current electric motor.

S

sack *n*: a container for cement, bentonite, ilmenite, barite, caustic, and so forth. Sacks (bags) contain the following amounts:

Cement	94 pounds (42.3 kilograms) (1 cubic foot)
Bentonite	100 pounds (45.5 kilograms)
Barite	100 pounds

salt mud *n*: 1. a drilling mud in which the water has an appreciable amount of salt (usually sodium or calcium chloride) dissolved in it. Also called saltwater mud or saline drilling fluid. 2. a mud with a resistivity less than or equal to the formation water resistivity.

salt water *n*: a water that contains a large quantity of salt, i.e., brine.

saltwater clay *n*: see *attapulgite*.

saltwater mud *n*: see *salt mud*.

sample mud *n*: drilling fluid formulated so that it will not alter the properties of the cuttings the fluid carries up the well.

samples *n pl*: 1. the well cuttings obtained at designated footage intervals during drilling. From an examination of these cuttings, the geologist determines the type of rock and formations being drilled and estimates oil and gas content. 2. small quantities of well fluids obtained for analysis.

sand *n*: in drilling, an abrasive material composed of small grains formed from the disintegration of any type of rock. Sand is larger than 74 microns (.074 mm or .003 inch) and smaller than 2,000 microns (2 millimetres, or 0.08 inch) in diameter.

sand content *n*: the insoluble abrasive solids contents of a drilling fluid rejected by a 200-mesh screen. Usually expressed as the percentage bulk volume of sand in a drilling fluid. See *mesh*.

sand trap *n*: a steel tank placed under the shale shaker into which mud falls after passing through the shale shaker. The shaker removes mainly cuttings from the mud so solids such as sand and other fine particles fall with the mud into the sand trap. Many of the solids fall out of the mud in the sand trap; those that do not settle out are removed with other specialized solids control mud treatment equipment such as desanders and desilters.

screen set *n*: an instrument for measuring the sand content of drilling mud.

seconds API *n*: the time in seconds that it takes 1 quart of drilling mud to flow out of a Marsh funnel. It is a measure of the mud's viscosity.

settling tank *n*: 1. the steel mud tank in which solid material in mud is allowed to settle out by gravity. It is used only in special situations today, for solids control equipment has superseded such a tank in most cases. Sometimes called a settling pit.

shaker *n*: see *shale shaker*.

shaker tank *n*: the mud tank adjacent to the shale shaker, usually the first tank into which mud flows after returning from the hole. Also called a shaker pit.

shale *n*: a fine-grained sedimentary rock composed mostly of consolidated clay or mud. Shale is the most frequently occurring sedimentary rock.

shale shaker *n*: a machine that uses one or more vibrating screens to remove cuttings from the circulating fluid in rotary drilling operations. Also called a shaker.

shale shaker screen *n*: a special wire mesh installed in a shale shaker that allows liquid mud to pass through but traps the cuttings and larger solid particles the mud carries from the hole.

shear *n*: action or stress that results from applied forces and that causes or tends to cause two adjoining portions of a substance or body to slide relative to each other in a direction parallel to their plane of contact.

shear stress *n*: force applied to a liquid to cause it to flow.

shear thinning *n*: a phenomenon in which the mud's viscosity decreases due to shear stress.

shock hose *n*: see *vibrator hose*.

shoulder *n*: a flat portion machined on a metal part that meets the shoulder of another part and serves to form a pressure-tight seal.

shut-in *adj*: shut off to prevent flow. Said of a well, plant, pump, and so forth, when valves are closed at both inlet and outlet.

SI *abbr*: International System of Units, or metric system.

silt *n*: a material that exhibits little or no swelling and whose particle size ranges from 2 to 74 microns. Dispersed clays and barite fall into this particle-size range.

single-acting *adj*: the action of a mud pump piston cylinder moving mud only on its forward stroke. Compare *double-acting*.

skid *n*: a low platform mounted on the bottom of equipment for ease of moving, hauling, or storing.

slough *v*: see *caving*.

slug *n*: a quantity of fluid injected into a reservoir to accomplish a specific purpose, such as chemical displacement of oil.

slug tank *n*: a relatively small separate tank or a small part of a larger tank that holds a small amount of mud for a special purpose. For example, it may hold a small quantity of heavy mud that will be used to slug the drill string. That is, the slug will be pumped into the string to keep mud from falling onto the rig floor when a drill pipe joint is broken out during a trip.

slug the pipe *v*: to pump a quantity of heavy mud into the drill pipe. Before hoisting drill pipe, it is desirable (if possible) to pump into its top section a quantity of heavy mud (a slug) that causes the level of the fluid to remain below the rig floor so that the crew members and the rig floor are not contaminated with the fluid when stands are broken out.

slurry *n*: a mixture in which solids are suspended in a liquid.

slush pit *n*: an old term for a mud pit. See *mud tank.*

slush pump *n*: see *mud pump.*

slush tank *n*: see *mud tank.*

sodium acid pyrophosphate (SAPP) *n*: a thinner used in combination with barite, caustic soda, and fresh water to form a plug and seal off a zone of lost circulation.

sodium hydroxide *n*: see *caustic soda.*

solid phase *n*: see *mud solids.*

solids *n pl*: see *mud solids.*

solids concentration *n*: total amount of solids in a drilling fluid as determined by distillation. Includes both the dissolved and the suspended or undissolved solids.

solids content *n*: see *solids concentration.*

specific gravity *n*: the ratio of the density, or weight, of a substance to the density of a reference substance. For liquids and solids, the reference substance is water.

spm *abbr*: strokes per minute.

spot *v*: to pump a designated quantity of a substance (such as acid or cement) into a specific interval in the well. For example, 10 barrels (1,590 litres) of diesel oil may be spotted around an area in the hole in which drill collars are stuck against the wall of the hole in an effort to free the collars.

spud in *v*: to begin drilling; to start the hole.

spud mud *n*: the fluid used when drilling starts at the surface, often a thick bentonite-lime slurry.

standpipe *n*: a vertical pipe rising along the side of the derrick or mast, which joins the discharge line leading from the mud pump to the rotary hose and through which mud is pumped into the hole.

still *n*: see *mud still.*

strainer *n*: a device placed upstream of a meter to remove foreign material from the stream that might damage the meter or interfere with its operation. The strainer element is generally coarser than a filter designed to remove solid contaminants.

stroke *n*: see *pump stroke.*

strokes per minute (spm) *n*: the number of times all the mud pump's pistons move forward and back to complete one stroke.

stuffing box *n*: a device that prevents leakage along a piston, rod, propeller shaft, or other moving part that passes through a hole in a cylinder or vessel. It consists of a box or chamber and a gland containing compressed packing.

suction *n*: in mud circulation, the drawing in of liquid into a pump.

suction dampener *n*: a pulsation dampener that is mounted on a mud pump's intake (suction) line. It absorbs the impact that occurs when the smooth flow of fluid in the suction line coming out of the suction tank meets the intermittent flow caused by fast-moving pump pistons.

suction head *n*: see *suction pressure.*

suction line *n*: the line that carries a product out of a tank to the suction side of the pumps. Also called the loading line.

suction pit *n*: also called a suction tank, sump pit, or mud suction pit. See *suction tank*.

suction pressure *n*: the pressure of a gas or fluid entering the suction valve of a compressor.

suction strainer *n*: a sieve-like device installed on the suction pipe of a mud pump that prevents large pieces of foreign material from being ingested by the pump.

suction tank *n*: the tank from which the mud pumps take suction. Also called a suction pit, sump pit, or mud suction pit.

suction valve *n*: on a mud pump, the intake valve through which mud is drawn into the pump.

sump pit *n*: see *suction pit.*

supercharging pump *n*: a centrifugal pump that ensures flooded suction to a mud pump by moving a large volume of mud from the suction pit into the mud pump suction.

surfactant *n*: a soluble compound that concentrates on the surface boundary between two substances such as oil and water and reduces the surface tension between the substances. The use of surfactants permits the thorough surface contact or mixing of substances that ordinarily remain separate. Surfactants are used in the petroleum industry as additives to drilling mud and to water during chemical flooding.

surfactant mud *n*: a drilling mud prepared by adding a surfactant to a water-base mud to change the colloidal state of the clay from that of complete dispersion to one of controlled flocculation. Such muds were originally designed for use in deep, high-temperature wells, but their many advantages (high chemical and thermal stability, minimum swelling effect on clay-bearing zones, lower plastic viscosity, and so on) extend their applicability.

surge dampener *n*: see *pulsation dampener*.

suspending agent *n*: an additive used to hold the fine clay and silt particles that sometimes remain after an acidizing treatment in suspension; i.e., it keeps them from settling out of the spent acid until it is circulated out.

suspension *n*: a mixture of small nonsettling particles of solid material within a gaseous or liquid medium.

swabbing effect *n*: a phenomenon characterized by formation fluids being pulled or swabbed into the wellbore when the drill stem and bit are pulled up the wellbore fast enough to reduce the hydrostatic pressure of the mud below the bit. If enough formation fluid is swabbed into the hole, a kick can result.

swivel *n*: a rotary tool that is hung from the rotary hook and traveling block to suspend the drill stem and to permit it to rotate freely. It also provides a connection for the rotary hose and a passageway for the flow of drilling fluid into the drill stem.

sx *abbr*: sacks; used in drilling and mud reports.

T

tattletale hole *n*: see *weephole*.

telltale hole *n*: see *weephole*.

ten-minute gel strength *n*: the measured 10-minute gel strength of a fluid is the maximum reading (deflection) taken from a direct-reading viscometer after the fluid has been quiescent for 10 minutes. The reading is reported in pounds/100 square feet. See *gel strength*.

thin *v*: to add a substance such as water or a chemical to drilling mud to reduce its viscosity.

thinning agent *n*: a special chemical or combination of chemicals that, when added to a drilling mud, reduces its viscosity.

thixotropy *n*: the property exhibited by a fluid that is in a liquid state when flowing and in a semisolid, gelled state when at rest. Most drilling fluids must be thixotropic so that cuttings will remain in suspension when circulation is stopped.

top drive *n*: a device similar to a power swivel that is used in place of the rotary table to turn the drill stem. It also includes power tongs. Modern top drives combine the elevator, the tongs, the swivel, and the hook. Even though the rotary table assembly is not used to rotate the drill stem and bit, the top-drive system retains it to provide a place to set the slips to suspend the drill stem when drilling stops.

torque *n*: the turning force that is applied to a shaft or other rotary mechanism to cause it to rotate or tend to do so. Torque is measured in foot-pounds, joules, newton-metres, and so forth.

toxicity *n*: the ability of a substance to be poisonous if inhaled, swallowed, absorbed, or introduced into the body through cuts or breaks in the skin.

Toxic Substance Control Act (TSCA) *n*: a U.S. congressional act that regulates the manufacture, processing, distribution in commerce, use, and disposal of chemical substances.

treat *v*: to subject a substance to a process or to a chemical to improve its quality or to remove a contaminant.

trip *n*: the operation of hoisting the drill stem from and returning it to the wellbore. *v*: shortened form of "make a trip."

trip in *v*: to go in the hole.

trip gas *n*: gas that enters the wellbore when the mud pump is shut down and pipe is being pulled from the wellbore. The gas may enter because of the reduction in bottomhole pressure when the pump is shut down, because of swabbing, or because of both.

triplex pump *n*: a reciprocating pump with three pistons or plungers. Triplex pumps are usually single-acting and are the most common type of mud pump today.

trip out *v*: to come out of the hole.

trip sheet *n*: a record of the measured drilling fluid that displaces the drill string as one pulls the pipe out of the hole or runs the pipe into the hole. See *trip tank*.

trip tank *n*: a small mud tank with a capacity of 10 to 15 barrels (1,590 to 2,385 litres), usually with 1-barrel or ½-barrel (159-litre or 79.5-litre) divisions, used to ascertain the amount of mud necessary to keep the wellbore full with the exact amount of mud that is displaced by drill pipe. When the bit comes out of the hole, a volume of mud equal to that which the drill pipe occupied while in the hole must be pumped into the hole to replace the pipe. When the bit goes back in the hole, the drill pipe displaces a certain amount of mud, and a trip tank can be used again to keep track of this volume.

trip tank indicator *n*: a device installed on a trip tank that shows the amount of mud being removed from or added to the trip tank.

U

underbalanced drilling *v*: to carry on drilling operations with a mud whose density is such that it exerts less pressure on bottom than the pressure in the formation while maintaining a seal (usually with a rotating head) to prevent the well fluids from blowing out under the rig. Drilling under pressure is advantageous in that the rate of penetration is relatively fast; however, the technique requires extreme caution.

underflow *n*: the lower discharge stream in a cone-shaped centrifuge. It moves downward and leaves by way of the apex.

V

vacuum *n*: 1. a space that is theoretically devoid of all matter and that exerts zero pressure. 2. a condition that exists in a system when pressure is reduced below atmospheric pressure.

vacuum degasser *n*: a device in which gas-cut mud is degassed by the action of a vacuum inside a tank. The gas-cut mud is pulled into the tank, the gas removed, and the gas-free mud discharged back into the mud tank.

valve *n*: a device used to control the rate of flow in a line to open or shut off a line completely, or to serve as an automatic or semiautomatic safety device. Those used extensively include the check valve, gate valve, globe valve, needle valve, plug valve, and pressure relief valve.

valve pot *n*: the machined tapered opening on the fluid-end body of a mud pump that holds the valve seat.

valve seat *n*: a tapered metal ring that seals against the valve pot and holds a valve in a mud pump.

velocity *n*: 1. speed. 2. the timed rate of linear motion.

venturi effect *n*: the drop in pressure resulting from the increased velocity of a fluid as it flows through a constricted section of a pipe.

venturi nozzle *n*: a nozzle that, because of the venturi effect, increases the velocity of a fluid flowing through it.

venturi tube *n*: a short tube with a calibrated constriction that is used in instruments or devices such as jet hoppers. It was developed to take advantage of the principle that a fluid flowing through a constriction has increased velocity and reduced pressure.

V-G meter *n*: see *Fan V-G™ meter, direct-indicating viscometer.*

vibrator hose *n*: short length of flexible hose that absorbs some of the vibrating pulsations that originate in a mud pump. Also called shock hose.

viscosity *n*: a measure of the resistance of a fluid to flow. Resistance is brought about by the internal friction resulting from the combined effects of cohesion and adhesion. The viscosity of petroleum products is commonly expressed in terms of the time required for a specific volume of the liquid to flow through a capillary tube of a specific size at a given temperature.

viscous *adj*: having a high resistance to flow.

volume *n*: the amount of a substance that occupies a particular space.

volumetric efficiency *n*: 1. actual volume of fluid put out by a pump, divided by the volume displaced by a piston or pistons (or other device) in the pump. Volumetric efficiency is usually expressed as a percentage. For example, if the pump pistons displace 300 cubic inches (4,916 cubic centimetres), but the pump puts out only 291 cubic inches (4,769 cubic centimetres) per stroke, then the volumetric efficiency of the pump is 97%. 2. for compressors, the ratio of the volume of gas actually delivered, corrected to suction temperature and pressure, to the piston displacement.

vortex *n*: 1. a whirling mass, e.g., a whirlpool. 2. in gas and liquid measurement, a rotational-flow zone that forms on alternate sides of a nonstreamlined body placed in a high-velocity flowing stream.

vortex finder *n*: the short pipe in a cone-shaped separator that extends down into the cone body from the top and that forces the whirling stream of material to start downward toward the small end of the cone body.

W

wait-and-weight method *n*: a well-killing method in which the well is shut in and the mud weight is raised the amount required to kill the well. The heavy mud is then circulated into the well while the kick fluids are circulated out. So called because one shuts the well in and waits for the mud to be weighted before circulation begins.

wall-building ability *n*: the ability of a drilling mud to plaster the wall of the hole with solids from the drilling mud.

wall cake *n*: also called filter cake or mud cake. See *filter cake*.

washout *n*: 1. excessive wellbore enlargement caused by solvent and erosional action of the drilling fluid. 2. a fluid-cut opening caused by fluid leakage.

water-back *v*: 1. to reduce the weight or density of a drilling mud by adding water. 2. to reduce the solids content of a mud by adding water.

water-base mud *n*: a drilling mud in which the continuous phase is water. In water-base muds, any additives are dispersed in the water. Compare *oil-base mud*.

water hammer *n*: see *fluid knock*.

water loss *n*: see *fluid loss*.

wear groove *n*: a groove in a metal part that when worn smooth indicates that the part should be replaced.

weephole *n*: a hole in the lantern ring of a liner packing in a mud pump. When the liner packing fails, fluid spurts out of the weephole with each stroke of the piston, indicating that the packing should be renewed. Also called telltale hole, tattletale hole.

weight *n*: see *mud weight.*

weighting material *n*: a material that has a high specific gravity and is used to increase the density of drilling fluids or cement slurries.

weight up *v*: to increase the weight or density of drilling fluid by adding weighting material.

wellbore *n*: a borehole; the hole drilled by the bit. A wellbore may have casing in it or it may be open (uncased); or part of it may be cased, and part of it may be open. Also called a borehole or hole.

wetting agent *n*: see *surfactant.*

whole mud *n*: all the components of the drilling mud, including reactive and nonreactive solids and the liquid phase (filtrate).

wildcat *n*: a well drilled in an area where no oil or gas production exists. *v*: to drill wildcat wells.

wiper *n*: a circular rubber device with a split in its side that is put around drill pipe to wipe or clean drilling mud off the outside of the pipe as the pipe is pulled from the hole.

Y

yield point *n*: 1. the maximum stress that a solid can withstand without undergoing permanent deformation either by plastic flow or by rupture. 2. in drilling mud engineering, another term for yield value.

yield value *n*: the resistance to initial flow, or the stress required to start fluid movement. This resistance is caused by electrical charges located on or near the surfaces of the particles. The values of the yield point and thixotropy, respectively, are measurements of the same fluid properties under dynamic and static states. The Bingham yield value, reported in pounds/100 square feet, is determined from a direct-indicating viscometer by subtracting the plastic viscosity from the 300-RPM READING. ALSO CALLED YIELD POINT.

Review Questions

LESSONS IN ROTARY DRILLING

Unit I, Lesson 7: Drilling Fluids, Mud Pumps, and Conditioning Equipment

True or False

Put a T for *true* or an F for *false* in the blank next to each statement.

_________ 1. A fluid can be a liquid or a gas but drilling fluids are only liquid.

_________ 2. The most common drilling fluids are based on water.

_________ 3. Most mud tanks are made from wood because wood swells when contacted by a fluid to form a leak-proof seal.

_________ 4. Usually, mud pumps are small units because they put out only about 200 horsepower.

_________ 5. Often, land rigs have at least two mud pumps; some offshore rigs may have two, three, or more mud pumps.

_________ 6. The annulus is the cylindrical space between the outside wall of the drill stem and the inside wall of the borehole.

_________ 7. A shale shaker's main job is to remove very fine drilled solids from drilling mud.

_________ 8. Desanders, desilters, mud cleaners, and mud centrifuges are designed to remove the coarsest (largest) cuttings from drilling mud.

_________ 9. A degasser is designed to remove entrained gas from drilling mud.

_________ 10. A kick is the undesirable entry of formation fluids into the wellbore.

Fill in the Blanks

Fill in the blanks with the appropriate word or phrase. Pick the correct term from those listed below.

at rest	viscosity
suspend cuttings	cuttings
lubricate and cool	stream or jet
surges	prevent
hydraulic power	support
swab	100 to 200 feet (30 to 60 metres)
wall cake	

Drilling fluid exits the bit in a high-velocity 11. __________________ or ______________ to move the cuttings away from the face of the formation. Liquid mud has to move from about 12. ___________ to ___________ per minute to lift cuttings. The 13. _______________

of a drilling mud is its resistance to flow. Gel strength is a measure of the mud's ability to 14. ______________ ________________. When pulling the drill string from the hole, crew members have to be careful not to 15. ______________ formation fluids into the hole. Similarly, when running the drill stem into the hole, they must be careful not to create 16. ____________ _____. One function of drilling fluid is to 17. _________________ and _______________ the bit and drill stem. Another function is to 18. __________________ the walls of the hole. Hydrostatic pressure is the force exerted by a fluid that is 19. __________________. A good 20. __________ _____________ slows the loss of liquid from the mud to a slow rate. Another important function of drilling mud is to 21. __________________ the undesirable entry of formation fluids into the wellbore. Drilling fluid can also transmit 22. _______________ _____ ____________ to operate downhole mud motors. 23. _______________ can reveal information about the formation being drilled.

Matching

Write the letter of the correct definition in the blank next to each term.

Terms

________ 24. trip tank

________ 25. pressure loss

________ 26. mud-return line

________ 27. mud tanks

________ 28. standpipe

________ 29. mud manifold

________ 30. reserve pit

________ 31. suction line

Definitions

a. Hold a supply of mud for the pump, receive mud circulated out of the well, store reserve mud, and provide a place for additives to be mixed into the mud

b. Holds a small quantity of mud used to keep track of the amount of mud the hole takes when crew members run pipe into and out of the hole

c. Connects two or more mud pumps to divide the flow of mud into several parts, combine several flows into one, or reroute a flow to different destinations

d. A pipe or a line through which mud from the suction tank goes to the mud pump

e. A rigid pipe that usually runs vertically up one leg of the derrick and which connects the mud pump's discharge and the top end of the rotary hose
f. A length of pipe through which drilling mud flows and which runs from the wellhead to the shale shakers
g. On some land rigs, a large earthen pit bulldozed out of the ground that holds waste fluids
h. The loss of mud pressure that occurs in the surface and subsurface components of the circulating system

Multiple Choice

Pick the *best* answer from the choices and place the letter of that answer in the blank provided.

_________32. Three phases of drilling mud are—
a. water, clay, and chemicals.
b. liquid, gas, and solids.
c. saltwater, clay, and reactive agents.
d. liquid, reactive solids, and nonreactive solids.

_________33. Water-base mud is mud that contains—
a. only fresh water.
b. only salt water.
c. either salt water or fresh water.
d. only oil.

_________34. An oil-base mud—
a. has more than 5% water.
b. consists of oil, emulsifiers, stabilizing agents, and salt.
c. is a water-base mud to which oil has been added.
d. does not perform well in high temperature wells.

_________35. Mud density (weight) is a measure of a mud's—
a. hydrostatic pressure.
b. viscosity.
c. wall building capabilities.
d. none of the above

_________36. A Marsh funnel measures a mud's—
a. weight (density).
b. resistance to flow.
c. hydrostatic pressure.
d. annular velocity.

_________37. To measure a mud's gel strength, a technician could use a—
a. Marsh funnel.
b. mud balance.
c. filter press.
d. direct-indicating viscometer.

_________38. Referring to Table 4 in the text, assume a rig is circulating a 10.5 ppg mud that needs to be weighted up to 12.0 ppg. The system contains 200 barrels of mud. How many sacks of barite have to be added to achieve this weight?
a. 96
b. 192
c. 384
d. 768

_________39. How much hydrostatic pressure does an 11.8-ppg drilling mud develop at the bottom of a 7,533-foot hole?
a. 46 psi
b. 462 psi
c. 4,622 psi
d. 46,222 psi

_________40. When handling caustic soda, crew members should—
a. add it to the mud system through the chemical barrel.
b. never add water to dry caustic soda.
c. wear appropriate personal protective equipment.
d. all of the above

True or False

Put a T for *true* or an F for *false* in the blank next to each statement.

_________41. For the most part, rigs use two types of pumps: reciprocating and centrifugal.

_________42. Today, most rigs use double-acting, duplex mud pumps.

_________43. A supercharging pump ensures that a mud pump's suction (intake) is flooded.

_________44. A discharge pulsation dampener creates pressure variations in a mud pump's discharge.

_________45. Sized, operated, and maintained properly, dampeners even out hydraulic power output, increase a pump's efficiency, and increase the life of the pump parts.

_________46. In general, duplex pumps are lighter in weight than triplex pumps.

_________47. Usually, it is not necessary to install a pressure-relief valve in a mud pump's discharge line because the discharge line, drill stem, or bit nozzles rarely become plugged.

_________48. It is usually not necessary to prime a mud pump after it has been shut down because pistons moving inside a dry pump liner are seldom harmed.

_________ 49. As long as the lubricating oil level in a mud pump's power end is correct, it is not necessary that it be clean.

_________ 50. A centrifugal pump uses pistons to move fluids.

Matching

Write the letter of the correct definition in the blank next to each term.

Terms

_________ 51. mud-gas separator
_________ 52. mud cleaner
_________ 53. pit-level indicator
_________ 54. flow sensor
_________ 55. desilter
_________ 56. jet hopper
_________ 57. centrifuge
_________ 58. shale shaker
_________ 59. differential flowmeter
_________ 60. hydrocyclone
_________ 61. desander
_________ 62. sand trap
_________ 63. degasser

Definitions

a. A device that has one or more vibrating or rotating screens
b. The tank that receives the mud after it first leaves the shale shaker
c. Removes solids from mud that range in size from 40 to less than 74 microns
d. Removes solids from mud that range in size from about 20 to 40 microns
e. Removes solids from mud down to about 7 microns in size
f. Removes solids from mud down to about 4 microns in size
g. A durable plastic or ceramic cone
h. A relatively small closed vessel or tank that removes large quantities of gas from mud
i. A steel tank that removes entrained gas from mud
j. A funnel-shaped device, usually connected to a centrifugal pump, that crew members use to mix dry or liquid materials to the mud system
k. Shows the level of mud in the mud tanks
l. Senses and indicates the rate of mud flow from the mud return line
m. Measures and indicates the difference between the rate of mud flow going in and coming out of the hole

Answers to Review Questions

LESSONS IN ROTARY DRILLING

Unit I, Lesson 7: Drilling Fluids, Mud Pumps, and Conditioning Equipment

True or false

1. F
2. T
3. F
4. F
5. T
6. T
7. F
8. F
9. T
10. T

Fill in the blanks

11. stream or jet
12. 100 to 200 feet (30 to 60 metres)
13. viscosity
14. suspend cuttings
15. swab
16. surges
17. lubricate and cool
18. support
19. at rest
20. wall cake
21. prevent
22. hydraulic power
23. cuttings

Matching

24. b
25. h
26. f
27. a
28. e
29. c
30. g
31. d

Multiple choice

32. d
33. c
34. b
35. a
36. b
37. d
38. c
39. c
40. d

True or false

41. T
42. F
43. T
44. F
45. T
46. F
47. F
48. F
49. F
50. F

Matching

51. h
52. e
53. k
54. l
55. d
56. j
57. f
58. a
59. m
60. g
61. c
62. b
63. i